Scientific Communication and National Security

A report prepared by the
Panel on Scientific Communication and National Security
Committee on Science, Engineering, and Public Policy

National Academy of Sciences
National Academy of Engineering
Institute of Medicine

NATIONAL ACADEMY PRESS
Washington, D.C. 1982

NOTICE: The National Academy of Sciences was established in 1863 by Act of Congress as a private, nonprofit, self-governing membership corporation for the furtherance of science and technology for the general welfare. The terms of its charter require the National Academy of Sciences to advise the federal government upon request within its fields of competence. Under this corporate charter, the National Academy of Engineering and the Institute of Medicine were established in 1964 and 1970, respectively.

The Committee on Science, Engineering, and Public Policy is a joint committee of the National Academy of Sciences, the National Academy of Engineering, and the Institute of Medicine. It includes members of the councils of all three bodies.

Library of Congress Catalog Card Number 82-62183

International Standard Book Number 0-309-03332-2

Available from

NATIONAL ACADEMY PRESS
2101 Constitution Avenue, N.W.
Washington, D.C. 20418

Printed in the United States of America

SPONSORS

This study was supported by the Department of Defense, the National Science Foundation, the American Association for the Advancement of Science, the American Chemical Society, the American Geophysical Union, and the National Academy of Sciences.

The NAS contribution was drawn from funds used for Academy-initiated projects; the funds were provided by the NAS consortium of private foundations. The consortium comprises the Carnegie Corporation of New York, the Charles E. Culpeper Foundation, the William and Flora Hewlett Foundation, the John D. and Catherine T. MacArthur Foundation, the Andrew W. Mellon Foundation, and the Rockefeller Foundation.

NATIONAL ACADEMY OF SCIENCES

OFFICE OF THE PRESIDENT
2101 CONSTITUTION AVENUE
WASHINGTON, D. C. 20418

September 30, 1982

Scientific Communication and National Security addresses one of the most difficult of policy issues: one in which fundamental national objectives seem to have been abruptly thrown into direct conflict. Advances in science and technology have traditionally thrived in an atmosphere of open communication; openness has contributed to American military and economic strength and has been a tenet of American culture and higher education. However, recent trends, including apparent increases in acquisition efforts by our adversaries, have raised serious concerns that openness may harm U.S. security by providing adversaries with militarily relevant technologies that can be directed against us. As would be expected when major national interests are in question, signs of distrust have appeared on all sides of the growing public discussion. The federal government, through its research and development agencies, and the university research community, where most basic research is conducted, both will lose much if the nation cannot find a policy course that reflects legitimate concerns.

The Panel on Scientific Communication and National Security was constituted to address this complex and critical issue. It combined unusual breadth, practical experience, and variety of viewpoint--from government, industry, and the scientific community. With energy and a sense of commitment, the Panel searched for a sensible and practical policy amid controversies that continued even as it carried out its deliberations. Chairman Dale Corson guided its systematic evaluation of the costs and the benefits of openness with patient wisdom.

The Panel has provided a set of principles that shows a way to resolve the current dilemma. However, the existence of valid principles is only part of what the nation needs; success in translating such ideas into practical governmental action is by no means assured. It is the Panel's hope and my own that it will be possible to establish within the government an appropriate group to develop mechanisms and guidelines in the cooperative spirit that the report itself displays. As the Panel points out, a key need is to improve mutual understanding between national security officials and members of the scientific community; representation of both in the process of implementing the report's recommendations would be an excellent first step.

Frank Press
President

PANEL ON SCIENTIFIC COMMUNICATION AND NATIONAL SECURITY

DALE R. CORSON (Chairman), President Emeritus, Cornell University
RICHARD C. ATKINSON, Chancellor, University of California, San Diego
  (Former Director, National Science Foundation)
JOHN M. DEUTCH, Dean of Science, Massachusetts Institute of Technology
  (Former Under Secretary, Department of Energy)
ROBERT H. DICKE, Einstein Professor of Physics, Princeton University
  (Former Member, National Science Board)
EDWARD L. GINZTON, Chairman of the Board, Varian Associates
MARY L. GOOD, Vice President and Director of Research, UOP,
  Incorporated (Member, National Science Board)
NORMAN HACKERMAN, President, Rice University (Former Chairman,
  National Science Board; Former Member, President's Science
  Advisory Committee)
JAMES R. KILLIAN, President Emeritus, Massachusetts Institute of
  Technology (Former Presidential Science Advisor; Former Member,
  President's Science Advisory Committee)
FRANKLIN LINDSAY, Chairman, Executive Committee, ITEK Corporation
RICHARD A. MESERVE, Attorney, Covington and Burling
WOLFGANG K. H. PANOFSKY, Director, Stanford Linear Accelerator Center,
  Stanford University (Former Member, President's Science Advisory
  Committee)
WILLIAM J. PERRY, Partner, Hambrecht and Quist (Former Under Secretary
  for Research and Engineering, Department of Defense)
SAMUEL C. PHILLIPS, Vice President and General Manager, TRW Energy
  Products Group (Former Director, National Security Agency)
ALEXANDER RICH, Sedgwick Professor of Biophysics, Massachusetts
  Institute of Technology (Former Member, National Science Board)
JOHN D. ROBERTS, Provost, California Institute of Technology (Member
  of the Council of the National Academy of Sciences)
HAROLD T. SHAPIRO, President, University of Michigan
CHARLES P. SLICHTER, Professor, Department of Physics, University of
  Illinois (Former Member, President's Science Advisory Committee;
  Member, National Science Board)
MICHAEL I. SOVERN, President, Columbia University
ELMER B. STAATS (Former U.S. Comptroller General; Former Deputy Budget
  Director; Former Executive Officer, Operations Coordinating Board,
  National Security Council)

Staff   LAWRENCE E. McCRAY, Project Director
        ELIZABETH G. PANOS, Administrative Assistant
        MITCHEL B. WALLERSTEIN, Staff Consultant

COMMITTEE ON SCIENCE, ENGINEERING, AND PUBLIC POLICY

GEORGE M. LOW (Chairman), President, Rensselaer Polytechnic Institute
SOLOMON J. BUCHSBAUM, Executive Vice President, Customer Systems, Bell Telephone Laboratories, Inc.
EMILIO Q. DADDARIO, Hedrick and Lane, Attorneys at Law
ELWOOD V. JENSEN, Professor and Director, Ben May Laboratory for Cancer Research, University of Chicago
ALEXANDER LEAF, Chief of Medical Sciences, Massachusetts General Hospital, and Jackson Professor of Clinical Medicine, Harvard Medical School
GARDNER LINDZEY, President and Director, Center for Advanced Study in the Behavioral Sciences
J. ROSS MACDONALD, William Rand Kenan, Jr., Professor of Physics, University of North Carolina
JOHN L. McLUCAS, President, World Systems Division, Communications Satellite Corporation
ELIZABETH C. MILLER, WARF Professor of Oncology, McArdle Laboratory for Cancer Research, University of Wisconsin
GEORGE E. PALADE, Chairman and Professor, Section of Cell Biology, Yale University School of Medicine
JOSEPH M. PETTIT, President, Georgia Institute of Technology
LEON T. SILVER, Professor of Geology, Division of Geological and Planetary Sciences, California Institute of Technology
HERBERT A. SIMON, Professor of Computer Science and Psychology, Carnegie-Mellon University
I. M. SINGER, Professor, Mathematics Department, University of California, Berkeley
F. KARL WILLENBROCK, Cecil H. Green Professor of Engineering, Southern Methodist University

Ex Officio

FRANK PRESS, President, National Academy of Sciences
COURTLAND D. PERKINS, President, National Academy of Engineering
FREDERICK C. ROBBINS, President, Institute of Medicine

ALLAN R. HOFFMAN, Executive Director
BARBARA DARR, Administrative Assistant

PREFACE

The use of American science and technology in the rapid increase in Soviet military strength over the past decade has aroused substantial concern in the current administration. This concern has been expressed frequently in recent months by high-ranking officials, who have called for tighter controls on all forms of technology transfer, including communication among scientists by such means as the publication of papers in scientific journals and by face-to-face meetings. In addition, federal agencies have already taken steps to control the flow of data and information from scientific research. These statements and actions have led to rising concern in the U.S. scientific community that such controls might impede scientific progress and its contribution to the national welfare.

In March 1982, discussions among officials of the Academy complex and the Department of Defense led to the creation of the Panel on Scientific Communication and National Security under the aegis of the Committee on Science, Engineering, and Public Policy, a standing committee, to study the question. The charge to the Panel was, generally, to examine the relation between scientific communication[1] and national security in light of the growing concern that foreign nations[2] are gaining military advantage from such research. It states four major elements, as follows:

- An examination of the national security interests and the interests in free communication in two or three specific fields of science and technology (e.g., cryptology,

---

[1] The Panel has concerned itself with scientific communication flowing from a range of research activities embracing basic and applied research and extending over a series of institutions, including universities, industrial laboratories, and government laboratories. A major share of the Panel's attention has been devoted to university research where no restraints on dissemination of findings--such as restraints to preserve proprietary interests, for example--have existed.

[2] The Panel has concentrated its effort primarily on the U.S.-U.S.S.R. relationship, given the level of concern about that problem and the limited time and resources available.

very high speed integrated circuits, artificial intelligence) to be selected by the study panel in consultation with the Department of Defense. This analysis will include an examination of the extent to which American research has been used in Soviet military programs and, if possible, a consideration of how such information was transferred. In addition, the Panel will assess and compare the contribution to Soviet military strength from the transfer of research information with that arising from other means of technology transfer, such as the Soviet acquisition of American hardware.

* A review--with an emphasis on the International Traffic in Arms Regulations (ITAR) and the Export Administration Regulations (EAR), and a proposed executive order on the classification system--of the principal policy and operational concerns of the respective government agencies, universities, scientific societies, and researchers. (The proprietary concerns of industry will not be considered.) The goal is to identify issues where common agreement exists, to expose those where apparent disagreements are based on misperceptions and misunderstandings, and, perhaps, to narrow and sharpen the issues on which genuine differences exist.

* A rigorous evaluation of critical issues concerning the application of controls on the flow of research information.

* The development of recommendations and conclusions concerning: (i) the intended and proper reach of controls vis-à-vis various categories of science and technology; (ii) areas of science and technology that are or should be outside the operation of controls; (iii) approaches that might provide more certainty and predictability to the regulatory system; and (iv) alternative procedures that might prove acceptable to all of the concerned sectors.

This study has been sponsored by the Department of Defense, the National Science Foundation, the American Association for the Advancement of Science, the American Chemical Society, the American Geophysical Union, and the National Academy of Sciences.[3] The Panel, composed of 19 members, includes senior members of university faculties and administrations, former federal agency officials, and leaders in high-technology industrial firms.

At the time the Panel was created, conversations among the Panel chairman, the President of the National Academy of Sciences, and the Under Secretary of Defense for Research and Engineering led to a decision that Panel members would be given security clearance (if they

---

[3] The NAS contribution was drawn from funds used for Academy-initiated projects; the funds were provided by the NAS consortium of private foundations. The consortium comprises the Carnegie Corporation of New York, the Charles E. Culpeper Foundation, the William and Flora Hewlett Foundation, the John D. and Catherine T. MacArthur Foundation, the Andrew W. Mellon Foundation, and the Rockefeller Foundation.

did not already possess it) so that it would be possible for them to receive classified information about technology transfers to other countries. The Panel was subsequently given three secret-level briefings by members of the intelligence community. In addition, a subpanel, comprising six members of the Panel who hold clearance at the highest level, was briefed at two additional meetings.

The Panel has examined the evidence provided at the intelligence briefings and has sought to deal with this information in a way that would eliminate the need to classify this report. The main thrust of the Panel's findings is completely reflected in this document. However, the Panel has also produced a classified version of the subpanel report based on the secret intelligence information it was given; this statement is available at the Academy to those with the appropriate security clearance.

The Panel invited as participants in its sessions liaison representatives from all the study's sponsors as well as from the departments of State and Commerce, the Office of Science and Technology Policy, the intelligence community, the Association of American Universities, the Institute of Electrical and Electronics Engineers, and the American Physical Society. Liaison members participated in the Panel's open sessions and those with the appropriate security clearance attended the Panel's classified briefings. A list of all those who participated in the Panel's deliberations is included (see pages 72-76).

The Panel held three two-day meetings in Washington at which it was briefed by representatives of the departments of Defense, State, and Commerce, and by representatives of the intelligence community, including the Central Intelligence Agency, the Federal Bureau of Investigation, the Defense Intelligence Agency, and the National Security Agency. The Panel also heard presentations by members of the research community and by university representatives. In addition to these briefings, the Rand Corporation prepared an independent analysis of the transfer of sensitive technology from the United States to the Soviet Union.[4] To determine the views of scientists and administrators at major research universities, the Panel asked a group of faculty members and administrative officials at Cornell University to prepare a paper incorporating their own views and those of counterparts at other universities (see Working Papers). The Panel also requested and received letters from a group of executives from high-technology industries expressing their views (see Appendix C). The Panel commissioned papers by experts in various aspects of technology transfer and studied the published material on the subject. It examined a few specific scientific areas in some detail.

In order to determine how and where controls might further the national welfare, it is necessary to balance many factors, including the military advantage from controls, their impact on the ability of

---

[4]This paper, among others, is included in the collected working papers used by the Panel. A photocopy is available from the National Academy Press, 2101 Constitution Avenue, N.W., Washington, D.C. 20418.

the research process to serve military, commercial and basic cultural goals, and their effects on the education of students in science and technology. The Panel hopes that this report serves to identify these important issues and to set out recommendations that achieve an appropriate balance.

The Panel is grateful for the assistance provided by the departments of Defense, State, and Commerce, and by the various intelligence agencies. Without their generous help, our task would have been impossible. The liaison representatives of the various departments, agencies, and organizations also contributed to our effort, and we thank them as well. We are also appreciative of the work of the Cornell University committee, which was headed by W. Donald Cooke. We wish to express special thanks to Frank Press, President of the National Academy of Sciences; Courtland Perkins, President of the National Academy of Engineering; and Philip M. Smith, Executive Officer of the National Academy of Sciences for their help and support. I wish to extend my personal thanks to Lawrence McCray, project director, Mitchel Wallerstein, staff consultant, and to Elizabeth Panos, administrative assistant, for their staff support. We are also grateful to Barbara Darr and Allan Hoffman of the COSEPUP staff. Finally, I wish to express my thanks to the individual members of the Panel for their dedicated service in making an early report possible.

                                                      Dale R. Corson
                                                      Chairman

CONTENTS

EXECUTIVE SUMMARY — 1
Unwanted Transfer of U.S. Technology, 1
Universities and Scientific Communication, 2
The Current Control System, 2
Costs and Benefits of Controls, 3
Principal Findings and Recommendations, 4

INTRODUCTION — 9

1 CURRENT KNOWLEDGE ABOUT UNWANTED TECHNOLOGY TRANSFER AND ITS MILITARY SIGNIFICANCE — 13
The Quality of the Evidence, 14
Potential Channels and Types of Technology Transfer, 14
The Soviet Acquisition Effort, 16
Evidence of the Extent of Unwanted Transfer, 17
Evidence of the Soviet Absorption Capacity, 18
Evidence of the Military Significance of Technology Losses, 19
Projections for Change, 19

2 UNIVERSITIES AND SCIENTIFIC COMMUNICATION — 22
University Research and Teaching, 22
Scientific Communication, 24

3 THE CURRENT CONTROL SYSTEM — 27
Classification of Information, 27
Export Controls, 28
Contractual Restrictions, 35
"Voluntary" Restrictions, 36
Controls on Foreign Visitors, 36

4 GENERAL CONCLUSIONS: BALANCING THE COSTS AND BENEFITS OF CONTROLS — 39
Preventing Soviet Military Advances Based on U.S. Research, 40
Fostering U.S. Military and Economic Strength, 42
Protecting Educational and Cultural Values, 45
The Feasibility of Controls, 46
Balancing Competing Objectives: The Panel's Judgment, 47

5 IMPROVING THE CURRENT SYSTEM   52
   The Workability of the Current System Can Be Improved, 53
   The Factual Basis for Decisions Can Be Improved, 57
   Better Mutual Accommodation Between the Government and
      Researchers Can and Must Be Achieved, 60
   U.S.-U.S.S.R. Scientific Exchange Programs Should Be Brought
      into Better Balance, 62

6 COMPILATION OF RECOMMENDATIONS   65
   Control of University Research Activities, 65
   The Workability of Export Controls on Scientific
      Communication, 66
   Data for Decision Making, 68
   The Government-University Relationship, 68
   U.S.-U.S.S.R. Scientific Exchanges, 69

ADDITIONAL COMMENT BY HAROLD T. SHAPIRO   71

LIST OF BRIEFERS, CONTRIBUTORS, AND LIAISON REPRESENTATIVES   72

LIST OF ACRONYMS   77

ANNOTATED BIBLIOGRAPHY   78

APPENDIXES

<u>Panel and Staff Papers</u>

A   Memorandum from the Intelligence Subpanel to the Panel
    on Scientific Communication and National Security
    (unclassified version)   91
B   The Historical Context of National Security Concerns
    About Science and Technology   97
    <u>Mitchel B. Wallerstein</u>
C   A Study of the Responses of Industry to a Letter of
    Inquiry from the NAS Panel on Scientific Communication
    and National Security   110
    <u>Edward L. Ginzton</u>
D   A Brief Analysis of University Research and Development
    Efforts Relating to National Security, 1940-1980   117
    <u>James R. Killian, Jr.</u>
E   Voluntary Restraints on Research with National Security
    Implications: The Case of Cryptography, 1975-1982   120
    <u>Mitchel B. Wallerstein</u>
F   The Role of Foreign Nationals Studying or Working in U.S.
    Universities and Other Sectors   126
    <u>Mitchel B. Wallerstein</u>

Background Documents

G	Letter from Five University Presidents	136
H	Statement of Admiral B. R. Inman for the May 11, 1982, Senate Governmental Affairs Subcommittee on Investigations Hearing on Technology Transfer	140
I	Executive Order on National Security Information	143
J	Correspondence Between the State Department and the University of Minnesota and M.I.T. Restricting Foreign Visitors	171

WORKING PAPERS OF THE PANEL

[Photocopies of the collected working papers of the Panel on Scientific Communication and National Security are available from the National Academy Press, 2101 Constitution Avenue, N.W., Washington, D.C. 20418.]

Soviet Science and Weapons Acquisition
   Arthur J. Alexander
Restrictions on Academic Research and the National Interest
   W. D. Cooke, Thomas Eisner, Thomas Everhart, Franklin A. Long, Dorothy Nelkin, Benjamin Widom, and Edward Wolf
East-West Technology Transfer
   John W. Kiser, III
Comments on Historical Aspects of Classification and Communication in Magnetic Fusion Research
   Richard F. Post, Melvin B. Gottlieb, and Wolfang K. H. Panofsky
The Office of Strategic Information (OSI), U.S. Department of Commerce, 1954-1957
   Mitchel B. Wallerstein
The Coordinating Committee for National Export Controls (COCOM)
   Mitchel B. Wallerstein with Annex by John P. Hardt and Kate S. Tomlinson

EXECUTIVE SUMMARY

The economic and military strength of the United States is based to a substantial degree on its superior achievements in science and technology and on its capacity to translate those achievements into products and processes that contribute to economic prosperity and national defense. There are concerns, however, that the Soviet Union has gained militarily from access to the results of U.S. scientific and technological efforts. Accordingly, there have been recent suggestions that tighter controls should be established on the transfer of information through open channels to the Soviets. Such controls would, however, also inhibit the free communication of scientific and technical information essential to our achievements. The Panel on Scientific Communication and National Security was asked to examine the various aspects of the application of controls to scientific communication and to suggest how to balance competing national objectives so as to best serve the general welfare. This task has involved a careful assessment of the sources of leakage, the nature of universities and scientific communication, the current systems of information control, and the several costs and benefits of controls. These assessments underlie the Panel's recommendations.

UNWANTED TRANSFER OF U.S. TECHNOLOGY

There has been a substantial transfer of U.S. technology--much of it directly relevant to military systems--to the Soviet Union from diverse sources. The Soviet science and technology intelligence effort has increased in recent years, including that directed at U.S. universities and scientific research. The Soviet Union is exploiting U.S.-U.S.S.R. exchange programs by giving intelligence assignments to some of its participating nationals. This has led to reports of abuses in which the activities of some Soviet bloc exchange visitors have clearly extended beyond their agreed fields of study and have included activities that are inappropriate for visiting scholars.

There is a strong consensus, however, that universities and open scientific communication have been the source of very little of this technology transfer problem. Although there is a net flow of scientific information from the United States to the Soviet Union, consistent with the generally more advanced status of U.S. science, there is serious doubt as to whether the Soviets can reap significant direct military benefits from this flow in the near term. Moreover, U.S. openness gives this nation access to Soviet science in many key areas, and scientific contacts yield useful insights into Soviet institutions and society.

UNIVERSITIES AND SCIENTIFIC COMMUNICATION

The principal mission of universities is education; in many American universities research has also become a major activity, but this research is intertwined with teaching and with the training of advanced research scientists and engineers. Participation in research teaches students to solve difficult, novel problems, often under the guidance of first-rate scientists. Federal policies in support of science have reinforced universities' dual functions.

The system as it has recently evolved has been remarkably successful; American research universities attract some of the best minds from around the world and are the principal source of our scientific preeminence. The effectiveness of this research is now seriously threatened, however, by a number of economic and social forces.

Scientific communication is traditionally open and international in character. Scientific advance depends on worldwide access to all the prior findings in a field--and, often, in seemingly unrelated fields--and on systematic critical review of findings by the world scientific community. In addition to open international publication, there are many informal types of essential scientific communication, including circulation of prepublication drafts, discussions at scientific meetings, special seminars, and personal communications.

THE CURRENT CONTROL SYSTEM

The government can restrict scientific communication in various ways. First, information bearing a particularly close relationship to national security may be subject to classification. This is the most stringent of the control systems because it serves to bar all unauthorized access.

Second, communications with foreign nationals may be restricted by export controls, such as those established by the Export Administration Act (EAA) and its associated Export Administration Regulations (EAR) and by the Arms Export Control Act and its associated International Traffic in Arms Regulations (ITAR).[1] Unless an exemption (or

---

[1] The Panel is aware that the Atomic Energy Act provides a unique statutory basis for controlling information bearing on nuclear weapons. The Invention Secrecy Act also allows patent applications to be kept secret for national security reasons.

"general license") applies, both systems require prior governmental approval for transfer of technical data--either in written or oral communication--to foreign nationals. Neither EAR nor ITAR is aimed at general scientific communication, and the Constitution limits the government's ability to restrain such communication. Nonetheless, some of the current discussion has focused on the application of export controls to scientific communication. This has proved particularly troubling to the research community in that the current control system appears to be vague in its reach, potentially disruptive, and hard to understand.

Third, the government can include controls on communications in the legal instrument defining the obligations of a recipient of government research funds. A proposal currently under consideration by the Department of Defense would require a DOD funding recipient to allow the government the opportunity for prepublication review of manuscripts dealing with certain research areas of national security concern.

Fourth, the government could attempt to influence conduct by seeking a voluntary agreement with researchers to limit the flow of technical information. Such an agreement is in place to enable the National Security Agency to review manuscripts dealing with cryptography and to negotiate alterations before publication.

Finally, communication with foreign nationals might be inhibited indirectly by limiting their access to the United States. The government can deny a visa request or impose restrictions on activities in this country. In addition, the government can directly regulate the admission of Soviet and East European visitors under particular scientific exchange agreements.

COSTS AND BENEFITS OF CONTROLS

Controls on scientific communications can be considered in the light of several national objectives. Controls can be seen to strengthen national security by preventing the use of American results to advance Soviet military strength. But they can also be seen to weaken both military and economic capacities by restricting the mutually beneficial interaction of scientific investigators, inhibiting the flow of research results into military and civilian technology, and lessening the capacity of universities to train advanced researchers. Finally, the imposition of such controls may well erode important educational and cultural values.

With respect to controls and Soviet military gains, the Panel notes that while overall a serious technology transfer problem exists, leakage from the research community has not represented a material danger relative to that from other sources. However, some university scientists will continue to expand their research beyond basic scientific investigations into the application of science to technologies with military relevance. This raises the possibility that the university campus will come to be viewed as a place providing much better opportunities for the illegal acquisition of technology. Information that is of special concern is the "know-how" that is gained by extended participation in U.S. research projects.

With respect to U.S. military and economic progress, controls may slow the rate of scientific advance and thus reduce the rate of technological innovation. Controls also impose economic costs for U.S. high-technology firms, which affect both their prices and their market share in international commerce. Controls may also limit university research and teaching in important areas of technology. The projected shortage of science and engineering talent can become the pacing factor in U.S. technological advance, so maintaining the flow of talented young people to military and commercial technology development efforts is particularly important. A national policy of security by accomplishment has much to recommend it over a policy of security by secrecy.

Apart from these considerations, the U.S. political system and culture are based on the principle of openness. Democracy demands an informed public, and this includes information on science and technology.

In addition, there are some inherent limits on the feasibility and effectiveness of controls. For example, controls cannot be expected to ensure long-term protection of sensitive information, given Soviet determination to procure data and the many parallel leakage channels, some of which are beyond U.S. jurisdiction. Finally, universities and most civilian research organizations lack the logistical capability to monitor the movement of information or personnel.

After weighing these benefits, costs, and feasibility assessments, the Panel arrived at a series of findings and recommendations.

PRINCIPAL FINDINGS AND RECOMMENDATIONS

Control of University Research Activities

The Panel found it possible to define three categories of university research. The first, and by far the largest share, are those activities in which the benefits of total openness overshadow their possible near-term military benefits to the Soviet Union. There are also those areas of research for which classification is clearly indicated. Between the two lies a a small "gray area" of research activities for which limited restrictions short of classification are appropriate.

The Panel's criteria leave narrow gray areas for which, in a few instances, limited restrictions short of classification are appropriate. An example of such a gray area may be a situation, anticipated in large-scale integrated circuit work, in which on-campus research merges directly into process technology with possible military application. In its recommendations the Panel has formulated provisions that might be applicable to such a situation.

All parties have an interest in having research work done by the most qualified individuals and institutions and in educating a new generation of capable scientists and engineers. These objectives must fit, however, within a system that enables the government to classify work under its sponsorship in accordance with the law and that enables the university to select only work compatible with its principal mission.

Unrestricted Areas of Research

The Panel recommends that no restriction of any kind limiting access or communication should be applied to any area of university research, be it basic or applied, unless it involves a technology meeting all the following criteria:

- The technology is developing rapidly, and the time from basic science to application is short;
- The technology has identifiable direct military applications; or it is dual-use and involves process or production-related techniques;
- Transfer of the technology would give the U.S.S.R. a significant near-term military benefit; and
- The U.S. is the only source of information about the technology, or other friendly nations that could also be the source have control systems as secure as ours.

Classification

The Panel recommends that if government-supported research demonstrably will lead to military products in a short time, classification should be considered. It should be noted that most universities will not undertake classified work, and some will undertake it only in off-campus facilities.

Gray Areas

The Panel recommends that in the limited number of instances in which all of the above four criteria are met but classification is unwarranted, the values of open science can be preserved and the needs of government can met by written agreements no more restrictive than the following:

    a. Prohibition of direct participation in government-supported research projects by nationals of designated foreign countries, with no attempt made to limit physical access to university space or facilities or enrollment in any classroom course of study. Where such prohibition has been imposed by visa or contractually agreed upon, it is not inappropriate for government-university contracts to permit the government to ask a university to report those instances coming to the university's attention in which the stipulated foreign nationals seek participation in any such activities, however supported. It is recognized that some universities will regard such reporting requests as objectionable. Such requests, however, should not require surveillance or monitoring of foreign nationals by the universities.

      b.   Submission of stipulated manuscripts simultaneously to the publisher and to the federal agency contract officer, with the federal agency then having 60 days to seek modifications in the manuscript. The review period is not intended to give the government the power to order changes: The right and freedom to publish remain with the university, as they do with all unclassified research. This does not, of course, detract from the government's ultimate power to classify in accordance with law any research it has supported.

The Panel recommends that in cases where the government places such restrictions on scientific communication through contracts or other written agreements, it should be obligated to record and tabulate the instances of those restrictions on a regular basis.

The provisions of EAR and ITAR should not be invoked to deal with gray areas in government-funded university research.

The Export of Domestically Available Technical Data Under ITAR and EAR Regulations

ITAR and EAR should be applied only where they can be effective, and then evenly to scientific communication from both universities and industry. Scientists have broad constitutional rights to disseminate information domestically and, as a practical matter, information that is available domestically is also available abroad.

It is the Panel's judgment that the national welfare, including national security, is best served by allowing the free flow of all scientific and technical information that is not directly and significantly connected with technology critical to national security. The Panel thus concludes that the government has the responsibility of defining in concrete terms those technical areas in which controls on information flow are warranted.

      1.   The Panel recommends that unclassified information that is available domestically should receive a general license (exemption) from the formal licensing process.

      2.   The Panel recommends that information that is not directly or significantly connected with technology critical to national security should also receive a general license (exemption) from the formal licensing process. The critical technology list approach--if carefully formulated--could serve to define those limited areas in which controls are appropriate.

The Use of Voluntary Controls

A system of voluntary controls has been inaugurated for prepublication review by the National Security Agency of manuscripts dealing with cryptography. The model established by this system may not be applicable to other areas because of the unique situation in the field of cryptography.

> The Panel concludes that the voluntary publication control mechanism developed for cryptography is unlikely to be applicable to other research areas that bear on national security. However, the Panel recommends that consideration be given to adopting this mechanism in future cases, if and where the appropriate preconditions exist.

The Militarily Critical Technologies List

The MCTL is drawn under congressional mandate for reference in export control administration. Part of the list is classified, thus denying its use to some potential "exporters" of data. Moreover, the list covers a wide span from specific items of hardware to generic definitions of technologies. The current list covers about 700 pages. As it stands, and also as the Panel understands the pending revision, this list is not a useful tool in guiding control of scientific or technical communication.

> The Panel recommends a drastic streamlining of the MCTL by reducing its overall size to concentrate on technologies that are truly critical to national security.

Technology Transfer to the Third World

The Panel has concentrated on the U.S.-U.S.S.R. relationship. However, there are clear problems in scientific communication and national security involving Third World countries. These problems in time might overshadow the Soviet dimension. This entire range of issues is both complex and important, and further intensive study is clearly indicated.

The Panel takes note of the current U.S. policy to help the People's Republic of China (PRC) advance its industrial technology. It is generally recognized that the capacity of the PRC to transfer such technologies to the military sector is limited. This technical assistance policy is not reflected, however, in restrictions the government is imposing on cooperative research and activities of PRC students at U.S. universities.

> The Panel notes that its deliberations did not extend to the complex issues raised by military-related technology

transfer from advanced industrial nations to Third World nations in regionally unstable areas or to those that may be potentially hostile to the United States and its allies. The Panel recommends that this subject receive further attention by the National Academy of Sciences or other qualified study groups under federal sponsorship.

INTRODUCTION

The military, political, and economic preeminence of the United States during the post-World War II era is based to a substantial degree on its superior rate of achievement in science and technology, as well as on its capacity to translate these achievements into products and processes that contribute to economic prosperity and the national defense. The success of the U.S. scientific enterprise has been facilitated by many factors, important among them the opportunity for American scientists and engineers to pursue their research--and to communicate with each other--in a free and open environment.

During the last two administrations, however, concern has arisen that the characteristically open U.S. scientific community has served as one of the channels through which critical information and know-how are flowing to the Soviet Union and to other potential adversary countries; openness in science is thus perceived to present short-term national security risks in addition to its longer-term national security benefits in improved U.S. military technology. Recent statements by senior administration officials have referred with alarm to the amount of information flowing from the United States to the Soviet Union and Eastern Europe. Deputy Secretary of Defense Frank Carlucci stated that

> in our considered view . . . the [scientific] exchanges to date, in the main, have not been reciprocal. Rather, it is quite apparent the Soviets exploit scientific exchanges as well as a variety of other means in a highly orchestrated, centrally directed effort aimed at gathering the technical information required to enhance their military posture ["Scientific Exchanges and U.S. National Security," Science, Vol. 215, January 8, 1982, p. 140].

This view has been expressed even more forcefully by Assistant Secretary of Commerce Lawrence J. Brady:

> Operating out of embassies, consulates, and so-called "business delegations," KGB operatives have blanketed the developed capitalist countries with a network that operates like a gigantic vacuum cleaner, sucking up formulas, patents, blueprints and know-how with frightening precision. We

believe these operations rank higher in priority even than the collection of military intelligence. . . . This network seeks to exploit the "soft underbelly"--the individuals who, out of idealism or greed, fall victim to intelligence schemes; our traditions of an open press and unrestricted access to knowledge; and finally, the desire of academia to jealously preserve its prerogatives as a community of scholars unencumbered by government regulation. Certainly, these freedoms provide the underpinning of the American way of life. It is time, however, to ask what price we must pay if we are unable to protect our secrets? ["Taking Back the Rope: Technology Transfer and U.S. Security," speech before the Association of Former Intelligence Officers, Washington, D.C., March 29, 1982, pp. 5-6].

The same dilemma was the focus of a speech by Admiral B. R. Inman, then Deputy Director of the Central Intelligence Agency, before the annual meeting of the American Association for the Advancement of Science:

There is an overlap between technological information and national security which inevitably produces tension. This tension results from the scientist's desire for unconstrained research and publication on the one hand, and the federal government's need to protect certain information from potential foreign adversaries who might use that information against this nation. Both are powerful forces. Thus, it should not be a surprise that finding a workable and just balance between them is quite difficult ["National Security and Technical Information," speech before AAAS, Washington, D.C., January 7, 1982, p. 1].

Why the recent concern? Administration officials and members of Congress began to question whether the nation's long-standing mechanisms for protecting militarily relevant secrets is still adequate, given the convergence of several independent recent trends in military technology (see Appendix B for the historical context of the current public debate).

Four perceived trends may explain the new sense of alarm. First, it is perceived that, at least in some important areas of military technology, the U.S. _lead_ over the Soviet Union is diminishing. Since American security in the post-World War II era has depended largely on technological superiority, the possible erosion of that edge is seen as significant. It is also perceived that--owing in part to the difficulties of nurturing scientific and technological growth in a closed society--the relative Soviet gains would not have been possible without the absorption of Western technologies. Those who take this view cite the high priority given by Eastern bloc intelligence services to the collection--by both overt and covert means--of scientific and technical information from the United States and its allies.

Second, it is argued that as military systems become more pervasively high-technology undertakings, the separation between military operations and scientific research is quickly narrowed. Along with technical sophistication--e.g., state-of-the-art guidance systems, lasers, improved cryptographic capabilities--comes the inevitable fact that scientists working at the research frontier are closer to military applications than they may have intended to be. Furthermore, various external factors seem to be pushing some universities deeper into applications-oriented research.

Third, a steadily increasing share of these technologies is dual-use in nature; that is, they have both military and nonmilitary applications. Much of the research in these dual-use areas is supported by commercial interests for exclusively nonmilitary purposes; examples include domestic microelectronics research, industrial robotics, and the developing interest in cryptographic research as a way to safeguard computer files. Two aspects of this perceived trend are significant: (1) in some areas of dual-use research, the government has lost its past monopoly on new knowledge--and its traditional leakage controls (classification and conditions written into research contracts) thus may have become insufficient, and (2) more and more researchers in the private academic and industrial communities who have been unaware of national security implications in their work find themselves confronted with potential restrictions on the dissemination of their findings.

Fourth, recent American foreign policy has had the effect of further expanding the already large number of potential leakage channels. East-West detente in the 1970s resulted in a significant expansion of trade with Communist nations, which has included technology agreements. The fear is that dual-use technologies may be inadvertently transferred in the process. In addition, scientific and other exchange programs with Eastern bloc nations and the People's Republic of China multiplied during the 1970s. Concerns about foreign abuse of these exchange programs for intelligence purposes also began to multiply, particularly in view of the concurrent perception that American universities were shifting toward research that is closer to technological frontiers.

The controls that had evolved earlier had concentrated on hardware and on technical information (e.g., troop movements, weapons manuals, blueprints) for which the national security implications were obvious. That system relied heavily on classification of documents and an export licensing system for physical products. The new trends that officials see in transfer of technology indicate a different focus--one that includes some scientific communications and some control of foreign scientific visitors. Much of the recent controversy can be interpreted as the result of government attempts to extend its controls to these new areas. For example, university researchers in microelectronics working under DOD contracts have been informed that dissemination of their results would be subject to existing export control regulations; permission for specific foreign scientific visits has been abruptly denied; papers have been withdrawn on short notice from international scientific meetings at government insistence;[1] consideration was given to removing the exemption for basic research in the executive order on classification; heightened enforcement efforts have detained foreign students returning home; and universities have been asked to help monitor and enforce restrictions on the movements of foreign scientists and foreign students on campus.

The current situation creates large dilemmas for U.S. policymakers. The U.S. military establishment wants to end the flow of militarily sensitive information to the Soviet Union, but finds that the controls available may also slow down the development of the United States' own military capabilities. This nation wants to keep its economy strong and to help other nations acquire know-how for their own economic growth, but there is fear that some of that know-how will later be turned to military ends that may endanger U.S. security.

In the heat of the current debate, it should be remembered that science contributes to several national goals that we all share, including the maintenance of U.S. military and economic strength, and a system of higher education and the pursuit of knowledge that serves as a world standard of excellence; U.S. research also may contribute to Soviet military strength. The Panel on Scientific Communication and National Security has attempted to make an initial net assessment of the extent and seriousness of U.S. technological losses, the effectiveness of present control mechanisms in dealing with the problem, and the costs of imposing controls on open scientific communication. In the process, the Panel has examined the problem from a broad range of perspectives in as comprehensive and objective a fashion as possible. It has sought to develop solutions that will provide maximum benefits, both in terms of maintaining the health of the U.S. scientific enterprise and safeguarding national security, while incurring minimum national costs.

---

[1] Since the Panel concluded its deliberations, the Department of Defense moved to prevent the oral presentation of unclassified papers at an international scientific meeting. Many researchers attending the 26th annual international technical symposium of the Society of Photo-Optical Engineers in San Diego, California, were informed only days before the session that their presentations might violate existing contractual obligations or export control regulations. Noting the presence of Russian and East European visitors at the symposium, DOD officials feared that scheduled papers--on such topics as optical technologies used in laser communications and infrared optics--would be of military significance. Government officials were also present at the meeting to personally warn speakers. In all, over 150 of the planned 626 scheduled papers were withdrawn. The incident has aroused confusion and controversy, particularly over the timing of the government's actions. Most of the papers that were withdrawn were to be presented by government employees or else involved work funded by DOD, raising the question of why timely review by funding officers had not been possible.

CURRENT KNOWLEDGE ABOUT UNWANTED TECHNOLOGY TRANSFER AND ITS MILITARY SIGNIFICANCE

This report is concerned with the benefits and costs of government controls on technology transfer, particularly as they apply to open scientific communication. To help gauge the potential impact of controls, the Panel on Scientific Communication and National Security undertook to gather available evidence on the extent and the military significance of past transfers.

The Panel took two steps to accomplish this objective. First, the Panel arranged extensive secret-level briefings (each followed by extensive open discussion) of the entire Panel by key spokesmen for the Defense Intelligence Agency, the Federal Bureau of Investigation, the Central Intelligence Agency, and the U.S. intelligence community's interagency Technology Transfer Intelligence Committee. Second, the Panel designated six members[1] to participate in more extensive all-source briefings. This subpanel set its own detailed agenda for these special briefings and reported its findings to the Panel. For the unclassified version of the subpanel report, see Appendix A. The original version of the report, classified at the secret level, is available at the Academy to those with appropriate clearance.

The Panel believes that it has obtained the existing evidence on the extent and nature of technology transfers--particularly those transfers that have involved or originated from the U.S. research community. While there has been extensive transfer of U.S. technology of direct military relevance to the Soviet Union from a variety of sources, there is a strong consensus that scientific communication, including that involving the university community, appears to have been a very small part of this transfer up to the present time. Open communication on basic research results, which is an essential part of this nation's open society and research process, has, however, contributed to the scientific knowledge base of the Soviet Union as well as to that of other nations.

---

[1] John Deutch, James Killian, Franklin Lindsay, Wolfgang Panofsky, Samuel Phillips, and Elmer Staats.

THE QUALITY OF THE EVIDENCE

The Panel's goal was to assess how much harm to our national security--in absolute terms and in relation to the larger problem--could be attributed to information losses from members of the scientific community, including university scientists.

The evidence on this question is incomplete for two basic reasons. First, the collection of data and analysis of the leakage problem have only recently begun. The interagency Technology Transfer Intelligence Committee, for example, was given its mission to examine the problem only in late 1981. The effort at present largely consists of the collection of information on incidents of technology transfer, and such data have not yet been organized in a way that would indicate the relative contributions of U.S. scientific sources or most of the many other sources of leakage. Second, the question is inherently much less tractable than most observers--on all sides of the policy debate--might wish. The development of a definitive answer as to the extent and significance of loss through any single channel of leakage would, in effect, require analysts to trace information from its origin within the United States, through a transfer channel to an adversary nation, and then into its use in a specific military application. This analysis would, in addition, have to extend to large enough numbers of specific instances to permit valid generalizations. Such a retrospective analysis would prove difficult enough if it were undertaken for ordinary domestic technology transfer, where all the principals in a transfer could be interviewed; meaningful analyses of international transfers, some involving extralegal means, are even more difficult. For the present, we are left with some indirect indicators and some individual case studies.

The technology transfer process can be seen as comprising four steps: (1) the _attempt_ by an adversary to obtain information, (2) the _actual transfer_ to the collecting nation, (3) the _absorption_ of the information into foreign technology, and (4) the resulting _improvement_ in the foreign country's military strength. Some parts of this transfer process--unfortunately, the parts that bear least directly on the Panel's ultimate question--are relatively well understood. Evidence of Eastern bloc attempts to secure Western technology, for example, is fairly extensive. Isolated occurrences of significant technology losses are fairly well documented, but none of these documented cases has involved open scientific communication. Evidence on the ability of the Soviet military to absorb Western technology is incomplete, while evidence on the military significance of identified transfers is largely fragmentary.

POTENTIAL CHANNELS AND TYPES OF TECHNOLOGY TRANSFER

The Overall Problem

The potential channels through which American technology may be lost are numerous and varied (see Table 1). The ability of a foreign

TABLE 1 Potential Technology Transfer Channels

Overt	Covert
1. Legal Direct Equipment Purchases	1. Illegal Direct Equipment Purchases
2. Legal Third-Country Purchases	2. Illegal Third-Country Diversions
3. Equipment Captured in Wars	3. Bribes to Western Nationals
4. Legal Licenses and Patents	4. Third-Country Visitors to United States
5. Turn-Key Plant Sales	5. Industrial Espionage
6. Joint Ventures	6. Foreign Agents
7. Direct Commercial Know-How	
8. Trade Shows, Exhibits, Conferences	
9. Academic Exchanges	
10. Open Literature, Including Government Publications	
11. Deliberate U.S. Government Leaks	

adversary to direct its acquisition effort to the least well policed and potentially most productive of these many channels is a major challenge to the United States, particularly when many channels are beyond the immediate control of the American government.

The Role of the Research Community

With respect to transfers involving the research community, it is useful to distinguish four types of information that may be of interest to foreign governments: (1) scientific _theory_; (2) knowledge of activities and _progress_ in specific scientific fields; (3) information that is embodied in scientific and technical _equipment_; and (4) experimentation and procedural _know-how_--detailed knowledge, much of it gained through direct observation and experience with scientific and technical techniques. The mechanisms for transferring these different types of information vary considerably. The first two types, for example, are transmitted in the written and oral messages commonly exchanged freely among researchers--not least in open publications. Equipment transfer involves the physical transportation of objects. The transfer of know-how involves information that is generally _not_ captured in scientific papers. The transfer mechanism for such detailed information involves neither documents nor equipment, but more typically is the "apprenticing" experience that takes place, among other means, through long-term scientific exchanges that involve actual participation in ongoing research. This last type of scientific communication is a leading concern of the U.S. intelligence community.

THE SOVIET ACQUISITION EFFORT

The Overall Problem

Systematic acquisition of Western technology has been a goal of Russian policy since well before the Russian Revolution of 1917. The current effort is pervasive, highly organized, dynamic, and well targeted. The effort is directed from the highest levels in the Soviet government; it involves Soviet intelligence services (e.g, the Committee for State Security--the KGB); the powerful Military-Industrial Commission (VPK), which is the central coordinating agency for all Soviet military R&D; the State Committee for Science and Technology (GKNT); and the Academy of Sciences (ASUSSR). The Soviets also make extensive use of satellite countries' intelligence networks, and targets are chosen from among the easiest sources. It should not be assumed that the extensive technology agreements between Eastern European countries and U.S. allies have no connection with the Soviet appetite for knowledge about U.S. technologies.

The GKNT oversees attempts to acquire knowledge about existing and new technologies from Western universities and high-technology firms and, if unable to do so by legal means, turns to clandestine means.

Most of the Soviet collection of scientific and technological information is performed by overt means. One impressive example of this overt effort is the employment of an enormous work force--involving tens of thousands of people--for the task of sifting and routing unclassified materials from around the world, including those published by the U.S. National Technical Information Service (NTIS).

Soviet science is part of the Soviet collection effort, to some extent reflecting a significant Soviet attempt in the last decade to bring Soviet science into the effort to foster military innovation. Soviet scientists and students who participate in formal international exchange programs have been linked to the intelligence effort.[2] One should assume that almost all Soviet technical visitors to the United States are prebriefed about specific acquisition needs, and it is certain that Soviet visitors to other countries are required to report on their foreign experiences. There is evidence that the quality of their reports is a possible factor in decisions about their future travel applications.

The KGB itself is known to have hundreds of "scientific officers" deployed throughout the world. In addition, a significant fraction of all Soviet scientific visitors are believed to have intelligence roles. The total number of visitors to the United States from other Warsaw Pact countries and Third World nations is much larger, and their U.S. travel is not controlled. Participation by some of them in Soviet collection efforts is certain.

---

[2] Soviet Acquisition of Western Technology, April 1982, pp. 1-5.

The Role of the Research Community

Officials in U.S. intelligence agencies have said that only a small fraction of the overall Soviet bloc intelligence collection effort is directed at U.S. universities. It can include individual students and scholars nominated to participate in exchange programs with the West who are routinely screened by Soviet intelligence agencies. Some Third World students who visit the United States are questioned by Soviet intelligence agents, and some may actually be recruited for intelligence roles. A recent trend in the collection effort aimed at the U.S. research community is an increased effort focusing on newly emerging technologies, particularly those that evolve directly from scientific research.

EVIDENCE OF THE EXTENT OF UNWANTED TRANSFER

The Overall Problem

Statements by intelligence community officials[3] indicate that about 70 percent of the militarily significant technology acquired by the Soviet Union has been acquired through Soviet and East European intelligence organizations, using both overt and covert methods. Most of the rest is acquired through legal purchases of equipment or data, publications, and through other Soviet organizations. The overall leakage is impressive; Table 2 shows examples provided by the intelligence community for just a single field, microelectronics.

The Role of the Research Community

Only "a small percentage" of the Soviet acquisition of militarily relevant information is said to come from communications involving scientists and students.[4] The Panel's inquiry revealed that specific evidence of such collections of information from U.S. sources almost always involved episodes in which visitor status was abused by Eastern bloc scientists. These reported activities cover incidents that did not clearly threaten U.S. security. Reported episodes have included cases in which (a) the visitor's technical activities and studies went beyond his or her agreed field of study; (b) the visitor's time was poorly accounted for, including reports of excessive time spent collecting information (e.g., in the library) not related to his or her field of study; (c) the visitor, either successfully or unsuccessfully, attempted to evade visa or exchange agreement restrictions imposed on

---

[3] Statement of Admiral B. R. Inman for the May 11, 1982, Senate Governmental Affairs Subcommittee on Investigations Hearing on Technology Transfer (see Appendix H).
[4] Statement of Admiral B. R. Inman, May 11, 1982 (see Appendix H).

TABLE 2 Microelectronic Equipment and Technology Legally and Illegally Acquired by the Soviet Bloc

Equipment or Technology	Comments
Process Technology for Microelectronic Wafer Preparation	The Soviets have acquired hundreds of specific pieces of equipment related to wafer preparation, including epitaxial growth furnaces, crystal pullers, rinsers/dryers, slicers, and lapping and polishing units.
Process Technology for Producing Circuit Masks	Many acquisitions in this area include computer-aided design software, pattern generators and compilers, digital plotters, photorepeaters, contact printers, mask comparators, electron-beam generators, and ion-milling equipment.
Equipment for Device Fabrication	Many hundreds of acquisitions in this area have provided the Soviets with mask aligners, diffusion furnaces, ion implanters, coaters, etchers, and photochemical process lines.
Assembly and Test Equipment	Hundreds of items of Western equipment, including scribers, bonders, probe testers, and final test equipment, have been acquired by the Soviets.

SOURCE: <u>Soviet Acquisition of Western Technology</u>, p. 9.

his or her itinerary; and (d) in one or two incidents a visitor participated in clearly illegal activities of an intelligence nature. The U.S. government, of course, is concerned about unwanted transfers even when visitors stay within their agreed-upon programs of study.

EVIDENCE OF THE SOVIET ABSORPTION CAPACITY

One should not necessarily equate foreign acquisition of sensitive technology with improvement of foreign military capabilities. Such improvements can occur only after an intermediate step is passed, namely, the successful exploitation of the acquired information.

Evidence of Soviet efficiency in absorbing Western technology is fragmentary and conflicting. On the one hand, there are indications of inhibiting tendencies in the Soviet system, including weak incentives to Soviet military designers to innovate, the fact that the acquisition effort is large and complex enough to assure inefficiencies in transferring information to those who have requested it, and the adverse effects of compartmentalization among Soviet scientists and engineers. On the other hand, the Soviet Union places a very high value on improving its military capabilities and may be able to overcome such impediments.

EVIDENCE OF THE MILITARY SIGNIFICANCE OF TECHNOLOGY LOSSES

The Overall Problem

There is no question that the overall loss of U.S. technologies from all sources to the Soviet Union has been extensive. The intelligence community has provided examples of Soviet acquisition of important technology (see Table 3).

The Panel has no reason to doubt government assertions that such acquisitions from the West have permitted the Soviet military to develop countermeasures to Western weapons, improve Soviet weapon performance, avoid hundreds of millions of dollars in R&D costs, and modernize critical sectors of Soviet military production.

The Role of the Research Community

With respect to the narrower question of losses associated with U.S. universities and other research organizations, discussions of the Panel with representatives of all U.S. intelligence agencies failed to reveal specific evidence of damage to U.S. national security caused by information obtained from U.S. academic sources. The reported episodes of abuses by Warsaw Pact visitors are disturbing, but they have not provided evidence of military consequences; the Panel's examination of the reported episodes did not reveal any resulting benefits to identifiable Soviet military systems. This negative finding is open to varying interpretations, given the incomplete and anecdotal quality of the existing evidence.

PROJECTIONS FOR CHANGE

While there is no indication that transfers of information involving the research community have, in the past, accounted for more than a very small part of the total leakage, the intelligence community believes that there is now a clear trend toward a greater Soviet effort to acquire information about technologies from universities and other research institutions. The factors cited to support this view are:

1. An increased Soviet emphasis in the past decade on the acquisition of newly emerging Western technologies.

TABLE 3 Selected Soviet and East European Legal and Illegal Acquisitions from the West Affecting Key Areas of Soviet Military Technology

Key Technology Area	Notable Success
Computers	Purchases and acquisitions of complete systems designs, concepts, hardware and software, including a wide variety of Western general purpose computers and minicomputers, for military applications.
Microelectronics	Complete industrial processes and semiconductor manufacturing equipment capable of meeting all Soviet military requirements, if acquisitions were combined.
Signal Processing	Acquisitions of processing equipment and know-how.
Manufacturing	Acquisitions of automatic and precision manufacturing equipment for electronics, materials, and optical and future laser weapons technology; acquisition of information on manufacturing technology related to weapons, ammunition, and aircraft parts, including turbine blades, computers, and electronic components; acquisition of machine tools for cutting large gears for ship propulsion systems.
Communications	Acquisitions of low-power, low-noise, high-sensitivity receivers.
Lasers	Acquisition of optical, pulsed power source, and other laser-related components, including special optical mirrors and mirror technology suitable for future laser weapons.
Guidance and Navigation	Acquisitions of marine and other navigation receivers, advanced inertial-guidance components, including miniature and laser gyros; acquisitions of missile guidance subsystems; acquisitions of precision machinery for ball-bearing production for missile and other applications; acquisition of missile test range instrumentation systems and documentation and precision cinetheodolites for collecting data critical to postflight ballistic missile analysis.
Structural Materials	Purchases and acquisitions of Western titanium alloys, welding equipment, and furnaces for producing titanium plate of large size applicable to submarine construction.
Propulsion	Missile technology; some ground propulsion technology (diesels, turbines, and rotaries); purchases and acquisitions of advanced jet engine fabrication technology and jet engine design information.
Acoustical Sensors	Acquisitions of underwater navigation and direction-finding equipment.
Electro-optical Sensors	Acquisition of information on satellite technology, laser rangefinders, and underwater low-light-level television cameras and systems for remote operation.
Radars	Acquisitions and exploitations of air defense radars and antenna designs for missile systems.

2. A belief that U.S. universities are expanding their participation in such research areas, particularly in process technologies.

3. A forecast that, as the government tightens its controls on other domestic sources of information and works with its allies to reduce third-country losses, foreign acquisition efforts will be increasingly redirected toward research institutions.

The Panel does not believe that it is yet possible to draw conclusions about this view. Even accepting the observation that Soviet collection efforts are being focused on the science underlying high-technology military applications, it does not follow that this is necessarily damaging to the United States. The concentration on basic science means that the military benefits to the Soviet Union would be long-range benefits that could become available to them from non-U.S. sources anyway. If the government succeeds in tightening the controls over loss mechanisms other than those associated with scientific communication in the United States, the loss through U.S. research institutions may become more significant. However, these other loss mechanisms are highly varied, and current Western control mechanisms, although improving, have far to go. Some of these inadequacies are structural, such as the limited membership and coverage of the Coordinating Committee (COCOM), an informal international organizaion for the coordination of national export controls; others have to do with the difficulties in preventing Soviet collection of information from nonaligned nations; and still others are due to limited resources and divided organizational responsibility.

For these reasons the Panel does not believe that a useful forecast can be made at present concerning the future proportion of leakage to the Soviet bloc through scientific communication.

UNIVERSITIES AND SCIENTIFIC COMMUNICATION

UNIVERSITY RESEARCH AND TEACHING

Universities are basically educational institutions, and this mission remains essential. American universities have also embraced research as a second principal mission since the latter part of the nineteenth century, and the two missions have since become highly interdependent, particularly at the graduate level.

It was not until the post-World War II era that the nurturing of basic research in U.S. universities, as supported by large-scale federal funding, became national policy. This was a deliberate decision that was based on the high productivity of universities--and university people--in support of national security needs during World War II. This experience led to support of basic research in universities by the Office of Naval Research immediately after the war, then to the creation of the National Science Foundation "to develop and encourage the pursuit of a national policy for the promotion of basic research and education in the sciences; [and] to initiate and support basic scientific research in the mathematical, physical, medical, biological, engineering, and other sciences . . ." (emphasis added).[1]

Today the research university is a major American institution, one that supplies almost all of the scientists and engineers for the academic, governmental, industrial, and military needs of the country and performs much of the fundamental research at the frontier of most important scientific fields. The research university is thus vital to the intellectual, economic, technological, and military health of the nation. There are 50--or perhaps 150, depending on how "major research university" is defined--of these institutions. While they account for less than 10 percent of the total R&D expenditures in the country, they account for more than half of the national effort in basic research, much of which is financed by the federal government. Thus, over the past 35 years the United States has moved deliberately to vest the primary responsibility for discovery of new knowledge--and for disseminating it--in a group of universities whose research activities are largely supported by the federal government. The success of this arrangement need not be argued.

---

[1] National Science Foundation Act of 1950, Sec. 3(a), 64 Stat. 149 (1950).

These institutions have created new technologies and indeed whole new fields of technology, as is illustrated by the new field of genetic engineering. They attract some of the ablest minds from all over the world. Their work is done in cosmopolitan groups, and while individual researchers may come and go, the research projects themselves continue in stable institutions that can assemble the best academic talent that exists worldwide in a given field.

Research and teaching have now become inextricably intertwined in the American research university. From the educational point of view, a critical mission of the university is to teach students how to solve difficult problems at the research frontier. Such research problems are difficult, because if they were not, they would already have been solved; they are novel or they would not be at the frontier. This method of education is carried out by apprenticing students to scientists who themselves are solving difficult problems at the research frontier. Undergraduate education is also strengthened when students are taught by scientists involved in discovering new knowledge firsthand.

Looked at from the point of view of research, the universities collect the ablest minds and provide them with an environment that gives them the freedom and resources to pursue their own ideas. An integral part of this environment is the postgraduate system whereby the leading students apprentice themselves to their mentors and provide the fresh outlook and energies characteristic of inquiring young minds. Such graduate programs generally conclude with a doctoral dissertation, independently conducted, which in itself is an original contribution at the research frontier.

Universities make several contributions to military and civilian technologies. Government agencies and private firms fund much university research to help solve technological problems and to get access to the best understanding that underlies such technologies. University researchers also occasionally provide insights by consulting directly with public- and private-sector R&D programs. Over the long term, however, the rate of technological advance in both sectors may be more seriously affected by the flow of new young talent--trained at the leading edge in relevant scientific disciplines--from universities to employment in military and industrial R&D efforts. The American university is the unique place in our society where new generations of leading scientists and engineers can be produced in sufficient numbers and proficiencies; in order to produce them the research function in the university mission must remain strong.

There are now, however, a number of economic, social, and political strains that, at the very least, will lead to significant changes in the way the system operates and, at worst, will lead to serious impairment of its effectiveness. Federal funding at universities, measured in constant dollars, leveled off about 15 years ago, and thus recent growth in the system has been slight, making it more difficult to replace obsolete equipment and to undertake new, and more expensive, enterprises. Demographic changes have led to a declining college-age population, which, with declining interest in science careers among young people, has raised questions about America's ability to attract adequate numbers of students to scientific and technical fields. Reduction in federal funds for the support of graduate students has exacerbated the problem. Economic considerations have led to falling numbers of engineering graduate students, leaving the country dependent on foreign nationals for a substantial fraction of its engineering faculty needs. Other problems involving the size and complexity of many research operations are also challenging and weakening the research-education system. Any additional challenges, including limitations on free communication, would compound an already difficult problem, making it yet harder to attract the best people to university research and making it harder for those who are attracted to maintain first-rate research programs.

SCIENTIFIC COMMUNICATION

Free communication among scientists is viewed as an essential factor in scientific advance. Such communication enables critical new findings or new theories to be readily and systematically subjected to the scrutiny of others and thereby verified or debunked. Moreover, because science is a cumulative activity--each scientist builds on the work of others--the free availability of information both provides the foundations for further scientific advance and prevents needlessly redundant work. Such communications also serve to stimulate creativity, both because scientists compete keenly for the respect of their peers by attempting to be first in publishing the answers to difficult problems and because communication can inspire new lines of investigation. Finally, free communication helps to build the necessary willingness to confront any idea, no matter how eccentric, and to assess it on its merits.

Scientific communication occurs in many different ways. Moreover, because no one country has a monopoly on scientific talent in any field, communication among research workers at the frontiers of science is international in character.

The most formal channel of communication is by means of publication of scientific findings in reputable journals. Over 2,000 such journals are widely distributed and universally read and cited by scientists. The international nature of science is reflected in the fact that in recent years only about 37 percent of the articles in these journals have been by U.S. authors. In fact, papers by U.S. scientists account for only 21 percent of the papers in chemistry and 30 percent of the papers in physics. Moreover, an analysis of the citations of articles in these journals shows that U.S. researchers make frequent use of foreign research results and, in fact, have increased their reliance on foreign results in recent years. For example, the West German chemical literature for 1979 received 20 percent more citations in U.S. literature than would be expected from examination of the West German share of the total literature in chemistry.[2]

---

[2] National Science Board, <u>Science Indicators 1980</u> (1981), pp. 16-18, 37-47, 222, 245.

Scientific meetings and symposia also play an important role in communication. Such meetings permit scientists to communicate their findings more rapidly than by publishing in a journal--and at the same time to receive instant feedback and ideas from their colleagues. The informal exchange of ideas that is characteristic of such meetings can also lead to significant modifications of research, to collaborative efforts, and to the avoidance of duplicative work. Because such meetings are most productive for all involved if the leading researchers in a given field participate, such meetings often attract international attendance.

Informal discussions among colleagues are also a critical element in scientific advance. Such communications obviously can and do occur most readily and frequently with colleagues in a researcher's own institution. For example, scientists in universities work closely with their own graduate students, and, as a result, graduate students are fully informed and totally immersed in the most advanced work in their fields. Such informal communications can also result in international transfers of information. Many graduate students in scientific and technical fields in U.S. universities are foreign students. (About 20 percent of the doctoral degrees from U.S. universities in 1979 were awarded to foreigners.) Moreover, it is common practice for preprints of research papers that will be published in scientific journals to be circulated among scientists working in the same field in the United States and abroad. There is also worldwide travel by major U.S. and foreign research workers who visit colleagues and their laboratories and exchange ideas.

There are also various governmentally sponsored international exchange programs in science and technology, including several with the Soviet Union, that are explicitly intended to foster international communication. These exchanges have figured prominently in the debate over national security losses.

Bilateral Intergovernmental Agreements

There have been bilateral (U.S.-U.S.S.R.) intergovernmental agreements in science and technology since 1972, when a total of 11 such agreements were signed by the two countries. These agreements covered a variety of joint programs in such fields as natural environment, space research, health, and oceanography.

These bilateral programs began to be reduced following the Soviet invasion of Afghanistan, and by 1980 the number of visits that took place under these exchanges had dropped by 75 percent. Four of the agreements were extended for a 5-year period in 1981, but following the recent events in Poland, the U.S. government decided not to renew the agreements that expire in 1982.

Certain fields covered by the bilaterals--for example, plasma physics, condensed-matter physics, and fundamental properties of matter--are areas of considerable Soviet strength. These exchanges have resulted in important contributions to science from both sides.

National Academy of Sciences Exchange Programs

Since 1959 the National Academy of Sciences (NAS) has operated an exchange program with the U.S.S.R. Academy of Sciences, providing for visits of from 1 to 12 months in duration for scientists and engineers in all fields. The Soviet participants have generally been older and more experienced than the American participants--visits to the United States are eagerly sought by Soviet scientists--but the quality of Soviet visitors has varied considerably. NAS has placed increased emphasis on assuring the professional competence of the Soviet visitors, but the Academy has had only limited success in obtaining specified visitors from the U.S.S.R. In recent years the size of the program has been reduced; the 1982 program level is 50 person-months of visits in each direction. In addition, periodic U.S.-Soviet symposia have taken place in various fields, such as radioastronomy, mathematics, and biochemistry. Although the quality of the meetings has been high, this type of meeting was suspended by NAS in 1980 in response to the Soviet treatment of Andrei Sakharov, who is a foreign member of NAS.

International Research and Exchange Board Program

The International Research and Exchange Board (IREX) administers a U.S.-Soviet exchange program under the sponsorship of the American Council of Learned Societies and the Social Science Research Council. Up to 50 Soviet graduate students and young faculty members participate each year, with visits lasting 9 months to a year. This program is conducted in cooperation with the U.S.S.R. Ministry of Higher and Specialized Secondary Education. Eighty to 90 percent of the U.S. participants have worked in the social sciences or humanities, while 90 percent of the Soviet participants have been involved in science or engineering. Despite this lack of symmetry, the program has served an important role in strengthening Soviet studies in the United States.

3

THE CURRENT CONTROL SYSTEM

There are five major instruments that the federal government can use to restrict scientific communication in the interest of national security. First, some communications are subject to the system for classifying and safeguarding information pertaining to national security. Second, communications with foreign nationals are subject to export controls under a variety of statutes. Third, scientific communications can be restricted through the legal instrument defining the obligations of recipients of federal funds. Fourth, communications can be restricted by voluntary agreement. Finally, communications with foreign nationals can be inhibited indirectly by limiting both foreigners' access to the United States and their activities while in this country. This chapter provides a brief overview of these control instruments.

CLASSIFICATION OF INFORMATION

Information can be classified for national security purposes under a program defined by executive order.[1] President Reagan has recently issued an executive order that, in broad outline, maintains the system established by earlier orders dating back to President Eisenhower (Exec. Order No. 12356, 47 Fed. Reg. 14874 (1982)).

The order states that a limited group of government officials (estimated by the White House at about 7,000 worldwide) has the authority to make an initial determination that certain information requires protection against unauthorized disclosure and to designate

---

[1] In addition, some particular categories of information are or can be classified by virtue of certain statutes. The Atomic Energy Act provides that information related to nuclear weapons and nuclear power is classified as soon as it comes into existence without the need for any governmental action (42 U.S.C. Section 2014(y), 2162). The Invention Secrecy Act allows the Commissioner of Patents to keep a patent application secret and to withhold the grant of a patent for national security reasons (35 U.S.C. Section 181).

the level of protection.² (Other officials who reproduce or extract classified information must apply the same classification that was in the original source material.) The system provides that such information may be designated as either top secret, secret, or confidential. Information is to be classified only if, at the least, unauthorized disclosure could be expected to damage national security. The categories of information eligible for classification include "scientific, technological, or economic matters relating to the national security" and "cryptology," but there is a specific exemption for "[b]asic scientific research information not clearly related to the national security" (Exec. Order No. 12356, 47 Fed. Reg. 14877 (1982)).

The government must have some preexisting connection with the information in order to classify it. Although the Reagan order deleted a provision in the previous order prohibiting the classification of research information that was not itself the fruit of access to classified information until the government had acquired a proprietary interest, the information subject to classification is still defined to include only that information that is "owned by, produced by or for, or is under the control of the United States Government" (Exec. Order No. 12356, 47 Fed. Reg. 14883 (1982)). The safeguarding requirement, which applies only to employees, contractors, licensees, or grantees, suggests the limits of governmental power.

A person is eligible for access to classified information only upon a determination both that the individual is trustworthy--a status that is customarily demonstrated by a security clearance at an appropriate level--and that access is essential to the accomplishment of lawful and authorized government activities. Each agency is required to establish a system to assure adequate protection of classified information, and a variety of statutes impose stringent penalties for wrongful behavior in connection with the information.

Classification is the most stringent of the five control systems because it serves to control all access to the information. The other systems of control are directed at communications with foreign nationals and, in some cases, only at communication through publication.

EXPORT CONTROLS

The chief controls on the export of technical data arise under the Export Administration Act (EAA) (50 U.S.C. App. Section 2401 et seq.) and the Arms Export Control Act (22 U.S.C. Section 2778).³ The EAA,

---

²The order requires "an employee, contractor, licensee, or grantee" who originates information that is believed to require classification to safeguard the information pending a classification determination by an authorized official.

³Other control systems may be important in particular situations. For example, regulations administered by the Nuclear Regulatory Commission or the Department of Energy govern the export of technology relating to nuclear equipment and materials.

which governs the export of articles or information with both military and civilian applications, is implemented by the Department of Commerce through a comprehensive set of regulations--the Export Administration Regulations (EAR) (15 C.F.R. Section 368.1-399.2). The Arms Export Control Act, which is directed at exports with a unique military function, is implemented by the Department of State through the International Traffic in Arms Regulations (ITAR) (22 C.F.R. Section 121.01-130.33). The Department of Defense and the intelligence community play an important role in the operation of both regulatory systems. Both EAR and ITAR can control the export of information that is unclassified.

In order to control the movement of militarily sensitive goods at the international level, the Coordinating Committee (COCOM) for national export controls was established by informal agreement in 1949. COCOM, which is composed of all the NATO countries (except Iceland) and Japan, has provided a forum for the consideration of trade controls on exports to the Warsaw Pact countries and the People's Republic of China. COCOM is a voluntary organization, and its decisions can only be implemented through the national policies of its members. These national policies, including the attitudes and approach to technical data exports, sometimes differ significantly. COCOM maintains three separate lists itemizing munitions, atomic energy, and dual-use items of particular concern, with the latter constituting the majority of the trade matters considered by the group. The United States has recently made particular efforts to strengthen the international control system and to bring about more uniformity and attention to the transfer of technical data as well as devices, but it has become evident that achieving significant changes will require a long-term, sustained effort.

EAA and EAR

The EAA authorizes the imposition of export controls for three principal reasons: to further national security, to foster foreign policy, or to protect the domestic economy from a drain of scarce materials (EAA Section 2402). The current debate on control of scientific information has focused on national security.

EAR sets out an elaborate system of licenses to control exports or reexports of tangible items to third countries. The stringency of government scrutiny depends on the nature of the article (EAR includes a Commodity Control List, which identifies the characteristics of goods of particular concern), the country of destination, and the end use of the goods.

The EAR controls also encompass "technical data" (EAR Part 379). An "export" of such data is deemed to occur whenever there is an actual transmission of data out of the United States, a release in the United

States with the knowledge that the data will be shipped out of the country, or a release abroad[4] (EAR Section 379.1(b)(1)).

With a few exceptions (e.g., certain exports to Canada), all exports of technical data must be made under either a general license (either a "General License GTDA" or a "General License GTDR") or a validated license[5] (EAR Section 379.2). A general license is analogous to an exemption: The license is extended automatically by force of regulation without either an application or the issuance of a document authorizing the export. A validated license, on the other hand, is a document authorizing a specific export. It is issued by the Office of Export Administration of the Department of Commerce following consideration of an application and letter of explanation from the exporter.

The General License GTDA, which is available for data exports to any destination, includes several categories of particular importance to scientists. First, the license is available for "data that have been made generally available to the public" through publications "that may be purchased without restrictions at a nominal cost or obtained without cost or are readily available at libraries open to the public" or through "open conferences" (EAR Section 379.3(a)). Second, the

---

[4] The regulations provide that a "release of technical data" may occur through:

(i) Visual inspection by foreign nationals of U.S. equipment and facilities;

(ii) Oral exchanges of information in the United States or abroad; and

(iii) The application to situations abroad of personal knowledge or technical experience acquired in the United States. [EAR Section 379.1(b)(2)]

[5] There are severe penalties for violations of the EAA or any regulation, order, or license issued under it. Certain willful violations are punishable criminally in the case of corporate entities by a fine of as much as $1 million or five times the value of the export involved, whichever is greater, or, in the case of an individual, by fine up to $250,000 or imprisonment for not more than 10 years, or both (EAA Section 2410(b)). Violations, whether or not willful, may also be subject to administrative penalties, including denial of export privileges, seizure and forfeiture of commodities or data, or, in certain cases, a civil penalty of up to $100,000 (EAA, Section 2410(c)). The imposition of administrative penalties is much more common than that of criminal penalties. Through "Project Exodus"--an aggressive effort of the Treasury Department's Customs Service in coordination with the Commerce Department--and through internal reforms, the overall enforcement activities have recently increased dramatically.

license is available for scientific data, which is defined to mean "information not directly and significantly related to design, production, or utilization in industrial processes" (EAR Section 379.3(b)(1)). Finally, the license is available for "instruction in academic institutions and academic laboratories, excluding information that involves research under contract related directly and significantly to design, production, or utilization in industrial processes" (EAR Section 379.3(b)(2)). These categories serve to limit the burden of the licensing system on most university scientists.

The availability of the General License GTDR is determined both by the destination of the export and the nature of the information. In its most significant application, this license permits the transfer of information concerning almost all nonmilitary industrial process technology throughout the free world. In certain cases the exporter must first receive written assurances that the data will not be reexported to certain other destinations (e.g., to the U.S.S.R.).

In FY 1981 the Office of Export Administration (OEA) processed more than 71,000 applications for validated licenses for exports, reexports, and the like and denied or returned without action about 9,000. Most of these applications were for the export of goods or data by industrial firms. OEA does on occasion have contact with university scientists, chiefly to ensure that a license is obtained before sensitive technology is transferred through exchanges with restricted-country foreign scholars. Since January 1982, OEA has sent 40 letters to host professors when OEA perceived the possibility of such a transfer. In all cases it is reported that steps were taken to ensure that the visitors would receive only information eligible for a General License GTDA.

It appears that there will be increasing emphasis in future years on controlling exports of technical data, with resulting changes in the control system. In part, this increasing emphasis will be the product of a growing concern in the intelligence community about such exports. In part, it will result from certain statutory changes.

In 1976 a Defense Science Board Task Force issued a report, commonly called the Bucy report,[6] suggesting that the export control system should shift from a focus on products to a focus on critical technology. Basically the Bucy task force argued that, with the exception of technologies of direct military value to potential adversaries, effort to control exports should not focus on the <u>products</u> of technology, but on <u>design</u> and <u>manufacturing</u> <u>know-how</u>. The report recommended that primary emphasis should be placed on (1) arrays of design and manufacturing know-how; (2) "keystone" manufacturing, inspection, and test equipment; and (3) products requiring sophisticated operation, application, or maintenance know-how. The Bucy task force concluded that the preservation of the U.S. lead in critical technological areas was becoming increasingly difficult but could be achieved, first, by denying the exportation of technology

---

[6] Defense Science Board Task Force on Export of U.S. Technology, <u>An Analysis of Export Control of U.S. Technology--A DOD Perspective</u> (Washington, D.C.: GPO, 1976).

when it represented a revolutionary (rather than evolutionary) advance for the receiving nation and, second, by strengthening the export control laws for critical technologies in the United States and in allied nations.

This change was accepted by the Department of Defense as official policy in 1977, and in 1979 Congress made it part of the EAA. Congress directed the Secretary of Defense to develop a militarily critical technologies list (the MCTL) and to incorporate this list, after review by the Department of Commerce, into the control system. DOD has worked for several years to develop and revise the MCTL, although it is not publicly available because portions of the list and the associated documentation remain classified. The list covers a broad spectrum of technologies, including many that have substantial or even primarily nonmilitary applications. Although it remains unclear how the MCTL will become part of a workable export scheme, it appears that the current government intention is to develop different and more sophisticated controls on data flows.

Arms Export Control Act and ITAR

The Arms Export Control Act and its associated regulations, ITAR, control the export of "defense articles and defense services." The controlled items are designated on the "United States Munitions List" (22 U.S.C. Section 2778). This list is divided into 22 categories, and for the most part the categories refer to military articles (e.g., firearms, tanks and military vehicles, military training equipment). The list also includes a category for "technical data" pertaining to the listed items.[7] Like EAR, ITAR includes an expansive definition of the term "export," including not only shipments from the United States but also disclosure of information during visits abroad by American citizens or disclosure to foreign nationals in the United States (ITAR Section 125.03).

Any export of technical data covered by the Munitions List requires prior approval and the issuance of a license by the Office of Munitions Control (OMC) in the Department of State, unless a specific exemption applies. Aside from an exemption for most data that are exported to Canada (ITAR Section 125.12), there are no distinctions among various export destinations. Moreover, the only exemption of significance for most scientific communications is one for data that are in published

---

[7]The definition of "technical data" in the ITAR includes not only data relating to items on the munitions list but also "any technology which advances the state-of-the-art or establishes a new art in an area of significant military applicability in the United States" (ITAR Section 125.01). However, the definition of technical data has been construed to refer only to data "significantly and directly related to specific articles on the Munitions List" so as to avoid interference with constitutionally protected speech (United States v. Edler Industries, Inc., 579 F.2d 516, 521 (9th Cir. 1978)).

form and subject to public dissemination (ITAR Section 125.11). In sum, although the scope of information subject to ITAR is narrower than that covered by EAR, the controls are more far-reaching. As in the case of EAR, both criminal and administrative penalties are available to punish violators (22 U.S.C. Section 2778(c); 22 C.F.R. Part 127).

In FY 1981 the OMC processed 35,800 license applications. As in the case of data provided by the Office of Export Administration, most of these applications were for the export of defense hardware items rather than data. OMC officials report that the number of cases involving university activities was "infinitesimal." Although the exact number could not be determined without a file search, they estimate that university-related applications might number 5 to 10 a year, at most. Whether wider academic awareness of ITAR requirements would multiply the frequency of these applications is not known.

Limitations of Export Control Authority

The most significant restraint on the government's use of the export control system to regulate communications is the First Amendment. The Supreme Court has made it clear that the First Amendment protects the right both to speak and to receive information and to communicate not only with other citizens but also with foreigners (see, e.g., Kleindienst v. Mandel, 408 U.S. 753, 762-765 (1972); Lamont v. Postmaster General, 381 U.S. 301 (1965)).[8] Scientific and technical communications generally appear to be entitled to full First Amendment protection.[9]

Before-the-fact restrictions on communication, such as those imposed by a licensing system, are the most serious and least tolerable limitations on First Amendment freedoms (see Nebraska Press Association v. Stuart, 427 U.S. 539 (1976); Southeastern Promotions, Ltd. v. Conrad, 420 U.S. 546 (1975)). This stems from the fact that, while subsequent punishment may chill speech and other communication, such prior restraints serve to freeze them entirely. The law thus reflects the view that a free society prefers to punish the few who abuse the rights of speech after they break the law rather than throttle them and all others beforehand. Moreover, even a system of subsequent punishment "requires the highest form of state interest to sustain its validity" (Smith v. Daily Mail Publishing Co., 443 U.S. 97, 102 (1979)).

Although the First Amendment protections are strong, they do not invalidate all governmental efforts to control the flow of technical data.

---

[8] The Court has not addressed the First Amendment rights of citizens outside our national boundaries. See Haig v. Agee, 453 U.S. 280, 308 (1981) (assuming arguendo that the First Amendment applies).
[9] See Ferguson, "Scientific and Technological Expression: A Problem in First Amendment Theory," Harvard Civil Rights--Civil Liberties Law Review 16 (1981):519.

First, if a speech is an integral part of a larger transaction involving conduct that the government is otherwise empowered to prohibit or regulate, the First Amendment does not give the speech immunity. A federal court considered this category of communication in United States v. Edler (579 F.2d 516 (9th Cir. 1978)) and upheld the application of ITAR on the ground that the President's authority to regulate arms traffic included "as a necessary incident" the authority to control the flow of information. In order to avoid constitutional difficulties, however, the court construed the ITAR to apply only to technical data that are "significantly and directly related" to specific articles on the Munitions List and, in instances where the information could have both peaceful and military applications, only to situations where the defendant knew or had reason to know that the information was received for a prohibited use (United States v. Edler, at 521).

Second, speech pertaining to commercial transactions is entitled only to limited First Amendment protection. Thus, the First Amendment might not prohibit controls on technical data that are disseminated for the purpose of promoting or proposing the sale of controlled items or related technical data.

Outside of these two areas the constitutional protection is broad, and this large residual category includes the scientific communications of greatest concern to the university community. Thus it might well be unconstitutional to use ITAR or EAR to bar an American scientist either from informing his or her colleagues, some of whom might be foreign nationals, of the results of an experiment or from publishing the results in a domestic journal.[10] Indeed, we understand that the ITAR and EAR are not currently used to restrict domestic publication, and the Office of Legal Counsel of the Justice Department has indicated in recent opinions that the application of the ITAR and EAR outside the two narrow areas might well be unconstitutional in many circumstances.

---

[10] Of course, restrictions on communication in particular cases might still be permissible. The Supreme Court has frequently indicated that protection of the national security could justify a prior restraint on speech. See, e.g., Near v. Minnesota (283 U.S. 697, 716 (1931)) ("No one would question but that a government might prevent [in time of war] . . . the publication of the sailing dates of transports or the number and location of troops"). But the Supreme Court has never approved such a restraint. Indeed, despite governmental assertions of strong national security interests, the Court explicitly refused to bar the publication of classified materials in the "Pentagon Papers" case (New York Times Co. v. United States, 403 U.S. 713 (1971)). Thus, although a district court recently enjoined the publication of an article that disclosed allegedly restricted technical data concerning the construction of a hydrogen bomb, United States v. Progressive, Inc., (467 F. Supp. 990 (W.D. Wis.)), appeal dismissed, (610 F.2d 819 (7th Cir. 1979)), in most circumstances a national security concern is unlikely to provide an adequate justification for a prior restraint.

CONTRACTUAL RESTRICTIONS

The federal government is a major source of funds for university research, and it can include controls on communications as one of the terms of the legal instrument defining the obligations of the funding recipient. Unlike the controls described above, such a system might be seen as voluntary in the sense that a recipient who disagreed with the contractual restrictions could choose not to accept federal support.

The Department of Defense is now considering a recommendation by a Defense Science Board Task Force[11] to establish a contractually based control mechanism for university research. At the time of contracting, the DOD project manager or contract monitor would negotiate with the university as to whether publications resulting from the research would be subject to restrictions. If so, the researcher would submit any papers to DOD for prepublication review to determine whether publication might result in the release of technical data of a type controlled under ITAR or EAR. The researcher would be allowed to publish if the government failed to act within a designated period (30 to 60 days) after the submission of a paper to it. But if the government concluded that the paper contained sensitive information, the researcher would be required either to modify the paper or to seek a license before publication.[12]

If DOD decides to adopt this recommendation, it has indicated that it might urge other funding agencies to use a similar approach and might assist other agencies both in developing guidelines for research contracts and in prepublication reviews.[13] The proposal would be a significant extension of the current export control system, because it would serve to restrict even domestic publication of unclassified data.[14]

---

[11] U.S. Department of Defense, <u>Report of the Defense Science Board Task Force on University Responsiveness</u> (Washington, D.C., January 1982).

[12] This recommendation may prove unacceptable to some universities.

[13] Statement by George P. Millburn, Acting Deputy Under Secretary of Defense for Research and Engineering (Research and Advanced Technology), to a Joint Session of the Science, Research and Technology Subcommittee and the Investigations and Oversight Subcommittee of the House Committee on Science and Technology, 97th Congress, 2d Sess. (March 29, 1982).

[14] Because the government is the funder of the research and because the funding recipient will have agreed beforehand to abide by the system of publication controls, the constitutional limitations on the government's authority under ITAR and EAR do not apply with the same force to the contractual control scheme.

"VOLUNTARY" RESTRICTIONS

The government has often found it possible to influence conduct by persuading the private sector to take certain actions voluntarily. Such voluntary controls--controls supported by no legal compulsion--might be applied similarly to restrain the flow of sensitive technical information. In fact, the National Security Agency (NSA) has begun a trial of a voluntary system through agreement with the Public Cryptology Study Group, which was established by the American Council on Education (see Appendix E). Researchers send papers with possible significance to the science of cryptology to NSA for prepublication review simultaneously with submission to a scientific journal. If NSA believes the paper contains sensitive information, an effort is made to negotiate modifications or postponement of publication with the author. To date, approximately 50 papers have been reviewed, and the few concerns raised by NSA were resolved amicably. The system provides that if negotiation is unsuccessful and the author remains dissatisfied, the paper may be submitted to a review panel.

CONTROLS ON FOREIGN VISITORS

Although control of the admission of aliens to the United States is obviously not a direct restraint on the flow of information, such control can significantly inhibit the flow by restraining both the interaction between domestic and foreign scientists and the observation of domestic equipment, data, and the like by foreigners.[15] Such controls might be established through either the visa process or, in the case of many scientific visitors from Eastern Europe, the implementation of particular exchange agreements.

Visa Controls

The admission of aliens to the United States is governed by the Immigration and Nationality Act (8. U.S.C. Section 1101 et seq).

---

[15]A similar restraint might be achieved by barring travel abroad by U.S. citizens. Although Congress has granted the President the power to restrict international travel by citizens in appropriate cases in the interests of national security, this power has very rarely been exercised (see Haig v. Agee, 97 S.Ct. 2766 (1981)). Moreover, the exercise of the power is subject to constitutional scrutiny (see Haig v. Agee; Kent v. Dulles, 357 U.S. 116 (1958)). Because an effort to control technical information flow by restraining international travel by U.S. citizens has not been attempted or even suggested, the sources and limitations of the government's authority will not be explored here. Disclosure of technical data by citizens who are abroad may require an export license, depending on the nature of the data and the persons to whom disclosure is made.

Typically, an alien must have a valid passport and must apply for admission and be granted a visa before being allowed to enter the United States. An alien is presumed to be an immigrant, and thus subject to strict admission controls, unless that person can establish that he or she is entitled to nonimmigrant status (8 U.S.C. Section 1184). Although there are widely varying grounds for admission as a nonimmigrant, those most relevant here are admissions:

* As a temporary visitor on business or pleasure;
* As a bonafide student pursuing a full course of study at an established institution of learning;
* As a visitor who is "of distinguished merit and ability" and who is coming temporarily to the United States "to perform services of an exceptional nature requiring such merit and ability";
* As a temporary visitor "who is a bonafide student, scholar, trainee, teacher, professor, research assistant, specialist, or leader in a field of specialized knowledge or skill, or other person of similar skill" and who is a participant in certain exchange programs designated by the Secretary of State.

(8 U.S.C. Section 1101(15)(B), (F), (H), (J)).

There are also numerous categories of aliens who are ineligible for admission (8 U.S.C. Section 1182(a)). For example, an alien may be excluded for various health reasons or for conviction of a crime involving moral turpitude. An alien also is ineligible if the consular officer or the attorney general knows or has reason to believe that the alien seeks to enter the United States "to engage in activities which would . . . endanger the welfare, safety, or security of the United States" or, under certain circumstances, if the alien is a member or affiliate of certain proscribed organizations. The burden is on the alien to prove eligibility to receive a visa[16] (22 C.F.R. Section 41.90). A nonimmigrant may be deported for failing to maintain the nonimmigrant status in which that person was admitted, for failing to comply with the conditions of admission, or for failing to avoid entry into certain of the ineligible categories (8 U.S.C. Section 1251(a)).

Judicial scrutiny of a refusal to admit an alien is exceedingly deferential. Control over admission is seen as an essential aspect of sovereignty, and thus the courts have recognized the plenary power of the Congress to make rules governing the matter. Moreover, so long as the Executive Branch acts within the boundaries of power delegated by the Congress, the courts will not interfere. However, the visa system is seldom used to inhibit technical communications, even for some Communist visitors, because of workload considerations and a lack of information about visa applicants and unresolved questions of policy as to whether visas can be denied for reasons of technology transfer.

---

[16] In addition, before issuing a nonimmigrant visa, the State Department may on occasion seek assurances from the host that he or she will seek to prevent the visitor's access to sensitive technologies at his or her research site.

Exchange Programs

Of particular relevance to the Panel is the class of foreign visitors admitted under international exchange programs that focus on science. Several of the exchange programs with the Soviet Union are described in Chapter 2. The Department of State, acting with advice from the Committee on Exchanges (COMEX), evaluates visitors from Communist countries under exchange programs.

4

GENERAL CONCLUSIONS:
BALANCING THE COSTS AND BENEFITS OF CONTROLS

The Panel attempted to identify and evaluate alternative approaches that simultaneously attain national security goals and cause the least damage to the capacity of the research community to make its many contributions to American life. Such a task required a review of the overall advantages and disadvantages--the benefits and costs--of the U.S. technology transfer control effort. This chapter lays out the general framework and principal results of the Panel's review.

The Panel has attempted to examine the relations between controls and three facets of the national interest: deterring advances in Soviet military strength that come about through the use of American research results; safeguarding continued progress in U.S. military and economic capabilities, which also depends in part on American research results; and protecting long-standing educational and cultural values. The Panel's observations about these relations in each of these areas are presented, together with observations on the feasibility of controls.

Although the Panel's mission was to investigate the effects of restrictions on scientific communication generally, it found in reaching its recommendations that the component of the American research community that requires separate consideration is the university. Restrictions on open communication have categorically different implications for universities than they do for industrial, governmental, and other components of the American research community. This is so for two primary reasons: first, universities alone integrate the research functions and degree educational programs, so that any adverse effects on research also adversely affect the quality of the next generation of scientists and engineers. Second, unlike other research institutions, universities have never established broad controls on access to ensure that sensitive proprietary or classified information is protected. Restrictions on communications thus present an unfamiliar and distinctly unwelcome challenge for university practices. Because the potential national security concerns are most likely to arise in work that is funded by the government, the Panel's conclusions concentrate on government-supported research.

Based on a review of costs and benefits, the Panel concludes (1) that most university research should be unrestricted; (2) that in rare cases such research should be classified for reasons of national security; and (3) that in a few specific cases, limited control measures short of outright classification may be warranted.

PREVENTING SOVIET MILITARY ADVANCES BASED ON U.S. RESEARCH

The Relation to Controls

The fundamental justification for controls is that they retard the rate of advance of Soviet military capacity by preventing Soviet access to relevant American science. Specific questions that must be evaluated in order to assess the merits of this justification are the extent to which Soviet military strength depends on U.S. technology in general, the extent to which Soviet military advances--either immediate or long-term--benefit from U.S. academic research, and the relative contribution to leakage accounted for by the different channels of scientific communication (e.g., visits by foreign scientists scientific visitors to the United States, published papers, oral presentations, and espionage).

The Panel's Assessment

The evidence reviewed by the Panel on the <u>overall problem of leakage</u> from all sources suggests that a substantial and serious technology transfer problem exists. A net flow of products, processes, and ideas is continually moving from the United States and its allies to the Soviet Union through both overt and covert means. A substantial portion of this unwanted transfer has been of little consequence to U.S. security, either because the United States did not enjoy a monopoly on a particular technology or because the technology in question had little or no military application. The Panel has also found, however, that a significant portion of the transfer has been damaging to national security.

The ease of global communications and the constant expansion of overseas sales of American products have increased greatly the number of points at which U.S. science and technology can be acquired. The loss of technology through non-U.S. sources continues to be a problem, despite efforts to reduce leakage from the West by tightening COCOM restrictions. In fact, as channels of leakage from the United States are closed or constricted, third countries may become much more attractive targets for acquisition. Stemming the overall flow of technology thus presents a very difficult challenge for the United States and its allies.

The losses of concern are not restricted to transfers to the Soviet Union. Technology transfers to Third World countries will permit them to modernize their military establishments faster and more efficiently; however, knowledge about the extent of this aspect of the leakage problem is fragmentary.

Leakage and the Research Community

The Panel took special care to assess the research community's contribution to the overall leakage problem. In its close examination of the cases that have involved a significant loss of technology that has been critical to national security, the Panel was shown no documented examples that were the direct result of open scientific communication. However, the absence of concrete evidence linking the research community with specific losses of information critical to national security does not imply a lack of Soviet intent to exploit this source. Moreover, a substantial volume of information has no doubt been transferred as a result of open scientific communication; information on basic research has contributed to the scientific base of the Soviet Union as well as to that of other nations.

Nonetheless, we are confident that fewer significant losses have occurred as a result of normal scientific communication than have occurred by other means. These other means include legal equipment purchases, outright espionage (particularly outside the United States), illegal conduct by some individuals and corporations in international trade, and secondary transfers through legal or illegal recipients abroad to the hands of U.S. adversaries.

The Panel observes that information acquired through open communication or by means of espionage activities on U.S. campuses may not often add substantially to the Soviet military capacity in the near term. The designers of Soviet military systems are conservative, and thus new scientific advances, whatever their origin, may not be readily adopted in military systems. Moreover, such information is probably best understood by Soviet researchers, and it may not flow readily to Soviet military designers because of the highly secretive and compartmentalized nature of the Soviet military R&D and procurement process.[1]

The Panel, therefore, concludes, based on all the above considerations, that, <u>in comparison with other channels of technology transfer, open scientific communication involving the research community does not present a material danger from near-term military implications</u>.

However, the present situation is dynamic and may call for future reevaluation. There has been a dramatic expansion during the past decade in the size and scope of the Soviet effort to obtain scientific and technological information, and this effort is much better organized and more carefully targeted today. The presence of Soviet intelligence agents in the United States--both on campus and elsewhere--is substantial. Also, some university scientists may continue to extend their research beyond basic scientific investigation into applications of technology with military relevance. This raises the distinct

---

[1] See Arthur J. Alexander, "Soviet Science and Weapons Acquisition," in the collected working papers for this report. Photocopies are available from the National Academy Press, 2101 Constitution Avenue, N.W., Washington, D.C. 20418.

possibility that the university campus will come to be viewed as a much better target of opportunity for the illegal acquisition of technology. There could be a convergence between the Soviet Union's enhanced acquisition capabilities and the possibilities of significant loss of technology through university campuses if certain types of research activities expand there in the future. An example of such research is found in some areas of microelectronics.

These areas of research are important because they develop know-how--practical techniques for solving technical problems--whose transfer to adversary nations could provide them with near-term military gains. Although published basic scientific results rarely have direct military application, technical know-how can be of substantial value in military design and production. In the Panel's view, it is at times as important to safeguard technical know-how in areas of rapid advance as it is to protect military systems themselves.

Special concern about the loss of know-how is justified in areas where process development, or "recipe" specification, is part of a research project. In these cases, research involves the development of a series of practical steps that makes it possible to use scientific principles to manufacture a product to satisfy production requirements. The recipe is therefore the critical know-how that must be protected, because the scientific concepts have little practical use without the manufacturing recipe.

Unlike the results of scientific research, know-how is rarely communicated effectively in written form or in brief conversations. The acquisition of know-how requires a long-term, hands-on working acquaintance with a scientific or technical area. Know-how is transmitted through on-campus apprenticeships--e.g., a visitor's continuous relations with a mentor and a research team in a research project--or through industrial training agreements. The importance of know-how thus raises special concern about the role of foreign visitors, including those in scientific exchange programs, who spend time working directly on high-technology projects likely to have near-term military applications.

Research that involves know-how is becoming more common in university laboratories for two reasons: (1) the development of equipment and processes for the manufacture of various items is often only an extension of the equipment and processes developed to conduct the basic research, and (2) as funding patterns and institutional relationships have changed, universities have sought more industrial support and have moved into areas closer to engineering design and product development than was generally the case in the recent past.

FOSTERING U.S. MILITARY AND ECONOMIC STRENGTH

The Relation to Controls

The fundamental question is the extent to which technology transfer controls intended to stem international leakage will also harm domestic communication and thus impede the contribution of American science to military and industrial advances. A second concern is whether controls on exchanges could reduce the ability of the U.S. government to understand Soviet capabilities and intentions.

In order to evaluate the potentially adverse effects of controls, it is necessary to assess the extent to which research contributes to improved U.S. military and economic performance and the extent to which controls that diminish the traditional openness of the scientific enterprise--thus limiting informal feedback, delaying the discovery of errors, narrowing critical evaluation, and complicating the scientist's search for predecessors' results--will diminish its overall contribution. In other terms, the question is whether controls intended to bar international communication will inevitably impede scientific communication within the United States. In considering the impact of controls on economic progress, the key question is whether restrictions increase firms' costs so as to affect the competitive position of American firms on the world market.

The Panel's Assessment

The Panel believes that scientific research and technological development are best nurtured in an environment where such efforts are dispersed but interdependent. Openness and a free flow of information are essential aspects of such an environment. The technological leadership of the United States is based in no small part on a scientific foundation whose vitality in turn depends on effective communication among scientists and between scientists and engineers. Thus, the short-term security achieved by restricting the flow of information is purchased at a price.

The contributions of basic research may be limited by any bars that would prevent researchers from learning from each other's work. If the research environment is altered in such a way as to discourage scientists from participating in certain areas of science, there will be fewer new ideas, and the pace of scientific advance will decrease. Openness, on the other hand, assures that new ideas are exposed to critical review by the best experts in the world. Only with such review can standards of U.S. research remain at high levels, and only such open review can ensure that logical errors, wrong paths, and unsupportable interpretations of data are avoided. Openness also fosters creativity, for it gives researchers assurance that they are building on the best and the newest ideas that exist worldwide. Prompt dissemination is also an enormous stimulus, since scientists are keenly competitive with one another. Moreover, openness leads with some regularity to the serendipitous answers to problems in one field through the use of research methods used for completely different purposes in other fields.

Openness and Military Strength

Restrictions on scientific communications may be costly because they could make work in those areas most relevant to U.S. military strength much less attractive to the best researchers and the best students. Both the rate of innovation in science-based technologies and the supply of young new technical talent trained at the frontier of knowledge in these fields could decline.

World War II demonstrated that the nation's top scientists and engineers--particularly those at the universities--constitute a major technical resource that can be of great value to our national security. The war showed decisively that military victory can depend on the outcome of the race between the scientists and engineers of adversary nations. Radar, for example, was developed just in time to play a critical role in the defense of Great Britain, and the subsequent development of microwave radar kept the allies ahead of other nations' countermeasures.

On a smaller scale, scientific communication through exchange programs can contribute to U.S. national security in several ways. There have been several cases in which Soviet scientists have made significant contributions to American research efforts. Furthermore, visits in both directions allow U.S. scientists to bring back informed appraisals of Soviet capabilities in science and technology (although not all scientific institutes in the Soviet Union are open to U.S. visitors). Exchange visits to the Soviet Union by U.S. scholars in fields other than science have great value in the development of Sovietology in this country. This has led to a better understanding of how Soviet society functions, which is essential to the formulation of well-informed U.S. policies.

Openness and Economic Strength

Restraints on scientific communication can limit the efficiency with which important information is transmitted to and within industrial firms.[2] The applications of new results from the research

---

[2] The Panel found that there is a general acceptance in private industry of restrictions on the technologies that are formally controlled under Defense Department contracts or by classification. However, leaders of U.S. technology-based industries believe that more stringent export controls, or even the continuation of current controls, raise the risk of restricting the flow of information within their firms and between Europe, Asia, and North America in a way that may ultimately be costly to American society. They believe that openness in scientific communication is vital to the economic vigor of the United States. These leaders also value the freedom to employ any personnel well trained in high-technology areas from the best universities. Many call for clearer identification of the technologies that should be restricted; clearer enunciation of the rationale for their control and of the damage that would occur without control; and means for ensuring prompt removal of controls when a technology is no longer at the state of the art or when the U.S. monopoly on information ceases through the normal process of international diffusion (see Appendix C).

community outside the universities will thus be impeded, possibly leading to higher internal research costs or to the loss of world market shares as the U.S. product performance falls behind that of foreign competitors. Particularly vulnerable are high-technology, dual-use technologies such as high-speed microelectronics, because these rapidly evolving areas could fall under national security controls that would inhibit the best researchers and advanced students from entering--and advancing--the field. Many high-technology firms depend on outside research scientists for contract research and consulting, and rely on universities to supply young talent trained at the frontier of technology.

A striking example of the economic benefit of openness is provided by the history of the transistor. The free dissemination of information about transistors in the early 1950s gave an extraordinary stimulus to the electronic and computer industries and led to the great U.S. superiority in these fields. This superiority has meant not only superior capacities in particular types of military and civilian equipment but also a strengthening of the economy as a whole.

Security by Accomplishment

To summarize, current proponents of stricter controls advocate a strategy of security through secrecy. In the view of the Panel security by accomplishment may have more to offer as a general national strategy. The long-term security of the United States depends in large part on its economic, technical, scientific, and intellectual vitality, which in turn depends on the vigorous research and development effort that openness helps to nurture.

A strategy of security by accomplishment has several institutional components. First, universities have the tasks of training new scientists and engineers and conducting basic research, the source of long-term progress. Second, government laboratories undertake research directed to particular national interests in defense, medicine, space energy, and agriculture. Third, industry translates the results of research into new commercial and defense technology. It is important that all these institutions attain their full potential, for economic as well as for military reasons. Open scientific communication plays an important part in keeping scientists and engineers in government, industry, and universities aware of each others' needs and findings.

PROTECTING EDUCATIONAL AND CULTURAL VALUES

The Relation to Controls

Controls on scientific communication could adversely affect U.S. research institutions and could be inconsistent with both the utilitarian and philosophical values of an open society. In order to investigate this potential effect, it is necessary to assess the extent to which controls on scientific communication disrupt the educational process--a process that has all but merged with the research function in major American research universities. With respect to cultural implications, it is necessary to assess the conflict between such restrictions and the health of our political system, the First Amendment's guarantees of free speech and a free press, and the need for an informed electorate.

The Panel's Assessment

American universities are particularly vulnerable to restrictions on scientific communication because of two special characteristics, namely, the critical role of research at many universities and the intimate relationship between university-based research and educational programs, particularly advanced training in research. Over the coming decades the nation's research performance will be heavily dependent on the continued health of university research programs, especially those in basic science.

Most university communities strive to maintain an "open society" where faculty and students freely share both the results of their research and the ideas that may develop into new areas of inquiry. In general, universities have seen such openness as an essential element of a scholarly environment. Controls on communication thus present a significant threat to a central tenet of university life and as a result are likely to discourage university-based scientists from participating in certain areas of science. Not only is the advance of science thereby slowed, but also the breadth of knowledge in the university community is thereby gradually diminished. Thus, the university can no longer fulfill its role as a central repository of knowledge.

With respect to cultural factors, the Panel believes that the costs of even a small advance toward government censorship in American society are high. The First Amendment's guarantee of free speech and a free press help account for the resiliency of the nation. If political authority is to be exercised effectively, there must be trust in government on the part of those affected--a trust that is promoted by openness and eroded by secrecy. Openness also makes possible the flow of information that is indispensable to the well-informed electorate essential for a healthy democracy. Openness also strengthens U.S. institutions by allowing comparison with the performance of others and nurturing adaptation to changed circumstances.

THE FEASIBILITY OF CONTROLS

The Panel believes it is important to keep a realistic perspective on the feasibility of controls. There is a danger that proponents of control measures may make the mistake of equating the _existence_ of controls with the lasting _denial_ of technologies to the Soviet military. Because the United States has a monopoly on only a fraction of all technology, and because the Soviet acquisition effort is carefully targeted, even totally effective U.S. controls would close but one channel among the many available to the Eastern bloc intelligence services. Efforts to restrict leakage by member nations of COCOM vary widely, and some industrially advanced nations, notably Sweden and Switzerland, are not members. Thus, in many cases losses can occur outside the United States and beyond the reach of its control efforts.

In addition, experience shows that it is difficult--if not impossible--to maintain total secrecy for long periods of time. Information leaks are inevitable--even when information is highly guarded--and others will, in time, make the same discoveries. The development of the atomic bomb provides an example of the latter assertion. Even without the the espionage attributed to Fuchs, Greenglass, and the Rosenbergs, the very use of the bomb revealed its basic character (through the production of intense radioactivity) and meant that trained scientists and engineers could easily make informed guesses about the bomb's nature from the known body of scientific theory. The reduction of theory to actual practice is by no means as easy, but this example does illustrate that secrecy can never replace the need to develop new ideas. Indeed, the most important fact about a technology is probably its very existence--not its design details. Once feasibility is proved in one country, other governments can confidently launch development efforts of their own. Concealing the existence of American military technologies is usually impractical.

The Panel also notes the inherent limitations in technology transfer controls as they might apply to research campuses. It is unclear how controls can be successfully applied to many activities that commonly take place on university campuses, including lectures, seminars, and discussions of faculty and students with visiting scientists. All of these are potentially subject to export regulations, yet all are, by long tradition, open activities. Moreover, any attempt to apply controls to such activities in a research environment (e.g., monitoring the movement of foreign scientists or students on campus) are logistically impractical, and therefore the chances of successful implementation are slight.

BALANCING COMPETING OBJECTIVES: THE PANEL'S JUDGMENT

After listening to the testimony, weighing the evidence, and pondering alternatives, the Panel concludes that the best way to ensure long-term national security lies in a strategy of security by accomplishment, and that an essential ingredient of technological accomplishment is open and free scientific communication. Such a policy involves risk because new scientific findings will inevitably be conveyed to U.S. adversaries. The Panel believes the risk is acceptable, however, because American industrial and military institutions have the capacity to develop new technology with a speed that will continue to give the United States a differential advantage over its military adversaries.[3]

---

[3] American industry has a greater ability to absorb new technology than does the industrial sector of the nation's adversaries, thereby

In any event, more than national security is at issue. Basic research investigations undertaken today may lead to applications in the long term (perhaps 10 to 20 years from now), often in unexpected ways. To attempt to restrict access to basic research would require casting a net of controls over wide areas of science that could be extremely damaging to overall scientific and economic advance as well as to military progress. The limited and uncertain benefits of such controls are, except in isolated areas, outweighed by the importance of scientific progress, which open communication accelerates, to the overall welfare of the nation. Security by accomplishment is a strategy that has served the nation well.

Principles for University Research

<u>The Panel concludes that the vast majority of university research, whether basic or applied, should be subject to no limitations on access or communications.</u>

Undoubtedly, some things must, by their very nature, be kept secret. It is clearly important, for example, to keep secret those properties of actual weapons systems that would enable a potential enemy to develop effective countermeasures. <u>Where specific information must perforce be kept secret, it should be classified strictly and guarded carefully</u>. The decision to accept or reject classified research projects or to establish off-campus classified facilities is a matter to be decided by universities.

The Panel concludes that <u>there are a few gray areas of research that are sensitive from a security standpoint, but where classification is not appropriate</u>. These research areas are at the ill-defined boundary between basic research and application and are characteristic of fields where the time from discovery to application is short. At present, a portion of the field of microelectronics is the most visible among the small handful of such new technologies.

---

providing the United States with a superior infrastructure in support of military technology. An important consideration here is the length of the military procurement cycle--the time from technological development to manufacture. Some have argued that the Soviet Union may be able to compensate to some degree, claiming it can introduce new technologies faster because it redesigns its military systems more frequently than the United States, providing an earlier opportunity to translate design into hardware. The Panel has not been able to explore this point, but it merits study.

Guidelines for Classified and Gray-Area Research

While it is impossible to specify classified and gray-area research with precision, there are some broad criteria that help to define the few areas in question.

<u>The Panel recommends that no restrictions of any kind limiting access or communication should be applied to any area of university research, be it basic or applied, unless it involves a technology meeting all of the following criteria:</u>

* <u>The technology is developing rapidly, and the time from basic science to application is short;</u>
* <u>The technology has identifiable direct military applications; or it is dual-use and involves process or production-related techniques;</u>
* <u>Transfer of the technology would give the U.S.S.R. a significant near-term military advantage; and</u>
* <u>The United States is the only source of information about the technology, or other friendly nations that could also be the source have control systems as secure as ours.</u>

In order to specify the areas where greater control would be appropriate, it may be useful to look at some examples of research that do <u>not</u> meet all of the above four criteria. Monoclonal antibody research is developing rapidly, and the interval from basic discovery to application may be short; but there appears to be no way in which this research could result in a significant military advance. Hence, there should be no need to impose controls in this field. Similarly, the science underlying aerodynamic design, even though it possesses obvious military significance, is a mature, slowly evolving field that is unlikely to provide any significant near-term military advantage to the Soviets. Thus, it too should be free of controls.

<u>The Panel recommends that if government-supported research demonstrably will lead to military products in a short time, classification should be considered. It should be noted that most universities will not undertake classified work, and some will undertake it only in off-campus facilities.</u>

In those few cases of government-sponsored research where national security considerations may require restrictions on publication, limitations on foreign access to facilities, or security classification, the Panel believes that certain guiding principles and procedures should be followed. <u>The provisions of EAR and ITAR should not be invoked to deal with gray areas in government-funded university research.</u> Rather, in the Panel's view, appropriate procedures should be incorporated in research contracts or other written agreements in those rare cases where some measure of control is required. The advantages of such provisions are that they give prior notice to the researcher that the funded research may turn out to have national security significance and foster a spirit of negotiated accommodation that helps prevent future misunderstandings about the researcher's obligations and recourse.

The Panel recommends that in the limited number of instances in which all of the above criteria are met but classification is unwarranted, the values of open science can be preserved and the needs of government can met by written agreements no more restrictive than the following:

a) *Prohibition of direct participation in government-supported research projects by nationals of designated foreign countries, with no attempt made to limit physical access to university space or facilities or enrollment in any classroom course of study. Moreover, where such prohibition has been imposed by visa or contractually agreed upon, it is not inappropriate for government-university contracts to permit the government to ask a university to report those instances coming to the university's attention in which the stipulated foreign nationals seek participation in any such activities, however supported. It is recognized that some universities will regard such reporting requests as objectionable. Such requests, however, should not require surveillance or monitoring of foreign nationals by the universities.*

Restrictions on access to nonclassified research, whether to research results or to physical facilities, are outside the normal operating procedures of research universities. It is, of course, within the power of the government to deny or issue conditional visas to foreign nationals who are believed to be seeking skills or technical data that will significantly damage our national security. In extraordinary circumstances, the government may seek to ensure that government-provided resources are not used to support nationals of specified countries who seek to work in specified programs. Access to program resources by nationals of designated foreign countries may be limited either through research contract terms or through other agreements negotiated with particular universities. Such contracts or agreements should not attempt to deny physical access to any university space or facility to any person accepted by the university into its community. The danger to national security lies in the immersion of a suspect visitor in a research program over an extended period, not in casual observation of equipment or research data.

b) *Submission of stipulated manuscripts simultaneously to the publisher and to the federal agency contract officer, with the federal agency then having 60 days to seek modifications in the manuscript. The review period is not intended to give the government the power to order changes: The right and freedom to publish remain with the university, as they do with all unclassified research. This does not, of course, detract from the government's ultimate power to classify in accordance with law any research it has supported.*

In some cases, a contractual agreement providing for simultaneous review of manuscripts at the time of their submission to scientific journals may be appropriate. A requirement for government comment within 60 days of submission of the manuscripts should provide adequate time for the government to assess the potential near-term military significance of the dissemination and to reach accommodation with the researcher before public release. Experience suggests that disagreements about publication can almost always be resolved by discussion between the principal investigator and the technical contract manager. The Panel emphasizes that its support for a review period is not intended to support any government effort to veto publication, or to limit the government's power to classify, in accordance with law, any research it has supported.

To help government policy officials to supervise the application of the gray-area research criteria and to gain perspective on the longer-term effects of the restrictions imposed on such research, there is a need to ensure that an accurate accounting of such restrictions is kept.

<u>The Panel recommends that in cases where the government places such restrictions on scientific communication through contracts or other written agreements, it should be obligated to record and tabulate the instances of those restrictions on a regular basis.</u>

5

IMPROVING THE CURRENT SYSTEM

In the previous chapter, the Panel defined a path concerning the control of university research that reflects our balancing of the competing national goals. In its assessment of the effectiveness and the costs of particular current and proposed controls, the Panel encountered several other areas in which adjustments can bring benefits and costs into better balance. This chapter presents specific ideas for improving the current system of controls.

The current system is undergoing rapid change. A few years ago the technology leakage problem was still largely defined in terms of hardware that could be copied or "reverse engineered," detailed blueprints, and so forth. Because the perceived nature of the technology leakage problem has shifted only recently, governmental control mechanisms themselves are still being adjusted. In a fundamental sense, the government is still in the early stages of a learning process as it reorients existing laws, policies, and programs--designed for other purposes--to achieve a new objective, the dimensions of which are still not fully limned. What makes this adjustment particularly difficult is that the current effort to understand and control unwanted technology transfer is, unavoidably, fractionated within the federal establishment. Four agencies, the Federal Bureau of Investigation, the Central Intelligence Agency, the Defense Intelligence Agency, and the National Security Agency, share the job of gathering intelligence on the nature, extent, and significance of unwanted transfers. Major regulatory authority is split among three separate offices (the Department of Commerce's EAR administrators, the Department of State's ITAR administrators, and State's visa-processing office). These offices depend heavily on outside units in the defense and intelligence communities for expertise as they reach their judgments. Similarly diffuse is the government's authority for classifying information and monitoring research and development results that it funds. Regulatory enforcement shows a similar diversity and includes yet another agency, the Department of the Treasury's Customs Service. The Panel discovered, not surprisingly, that few people either inside or outside the government have a comprehensive understanding of the government's technology transfer control effort. Individual components are understood, but their relations to one another--and, significantly, their implications for scientific productivity as these programs shift to address the scientific basis of technologies--are only now beginning to emerge.

In general, the Panel concludes that <u>there is much room for improvement in intelligently targeting the government's efforts to prevent unwanted technology transfer; priorities must be set and communicated in order to limit the adverse effects of controls on other vital national interests, including that of maintaining a position of world leadership in science</u>.

More specifically, there are several areas in which improvement is needed:

* making controls more <u>workable</u>,
* improving the <u>factual basis</u> for decisions,
* improving <u>mutual understanding</u> between the government and the scientific community, and
* bringing better balance to U.S.-U.S.S.R. <u>exchange programs</u>.

THE WORKABILITY OF THE CURRENT SYSTEM CAN BE IMPROVED

At this stage, the government's technology transfer controls have a very wide compass. There is a risk that by covering such broad expanses of technology and by implicating the scientific activities across the same range, the overall efficiency of the effort will suffer. Two principles can be applied to bring a more coherent focus to the problem. First, the government should concentrate on the most <u>feasible forms of control</u> and should eschew regulations that impose compliance burdens without significantly affecting leakage. Second, it should concentrate its resources more systematically on those <u>technologies that are of greatest relevance</u> to near-term Soviet military strength.

Export Controls and Domestically Available Information

American scientists have broad, constitutionally based rights to disseminate information within the United States free from government control, unless the information is classified or they have agreed in advance to contractual provisions limiting disclosure. And, as a practical matter, information that is available domestically is also available abroad. For example, there is no practical way to prevent domestic publications from circulating internationally. Both ITAR and EAR recognize this fact to a limited extent by providing exemptions from the formal licensing process for certain types of generally available information, such as published data.[1] But information is available through many channels--lectures, seminars, conferences,

---

[1] The Atomic Energy Act provides a unique statutory basis for controlling information bearing on nuclear weapons.

lecture notes, and the like. In recognition of the impossibility of impeding the flow of such domestically available information, control systems should be focused elsewhere.

The Panel recognizes the existence of the hypothetical danger that when research is not done under a government contract prohibiting disclosure, a broad exemption might facilitate dissemination of work of national security concern. In practice, however, other safeguards already minimize this risk. First, research supported by industry often has proprietary value, and researchers are unlikely to release it to competitors, international or domestic. Second, most research of near-term military relevance is done under government contract, already giving the government an opportunity to classify or reach direct agreement about restrictions on dissemination. Third, the Panel believes that researchers, if aware of the government's bona fide national security concerns, would be responsive. Thus, in light of the safeguards and the impracticality of control, a broad exemption is appropriate.

<u>The Panel recommends that unclassified information that is available domestically should receive a general license (exemption) from the formal licensing process.</u>

Priorities Within the Export Control System

It is the Panel's judgment that the national welfare is best served by allowing the free flow of all scientific and technical information that is not directly and significantly connected with technology critical to national security. Any diffusion of effort to control such information brings difficulties for administrators of the export control system because they must spread their resources across many technical areas. It also raises researchers' fears of potential vulnerability to government enforcement actions in fields that are far removed from national security concerns.

<u>The Panel recommends that information that is not directly and significantly connected with technology critical to the national security should also receive a general license (exemption) from the formal licensing process. The critical technology list approach—if carefully formulated—could serve to define those limited areas where controls are appropriate.</u>

Militarily Critical Technologies List

The Militarily Critical Technologies List (MCTL) was originally seen as a way to help shift the emphasis in export controls away from products toward the control of know-how, as had been recommended by the Bucy report in 1976. There is a real danger that the pending MCTL, if applied to scientific communications, will serve to make the export control effort more diffuse rather than to help the government focus on the most critical areas of concern. The current version is stretched too broadly among a long (700 pages) and expansive array of technologies, all defined as militarily critical. This hampers any agency efforts to use the list as an effective the basis for licensing, monitoring, or enforcement.

The Panel is concerned about the tendency to expand the MCTL, exacerbating the problems of understanding and applying it. For example, among the critical technologies listed in the most recent draft of the MCTL are techniques for volume production of microwave tubes, techniques for fabrication of multigap solar cells, and rare earth magnet materials such as samarium cobalt. Although such techniques and materials are essential in some military applications, they are also commonly employed in ordinary commercial processes in several parts of the world. If such a list were applied literally, it would appear that much basic research would be subject to export controls.

As the MCTL now stands, and as the Panel understands the current revision, it is too unwieldy to be useful in guiding government controls of scientific and technical communication.

The Panel recommends a drastic streamlining of the MCTL by reducing its overall size to concentrate on technologies that are truly critical to national security. The Panel recommends that items should be removed from the MCTL if they are in one or more of the following categories:

    a. science and technology whose transfer would not lead to a significant near-term improvement in Soviet defense capability;

    b. science underlying a mature technology—that is, a technology that is evolving slowly;

    c. science underlying dual-use technology that is not process-oriented;

    d. components used in militarily sensitive devices that in themselves are not sensitive.

The Panel recognizes that technology transfer controls may be adopted for reasons other than direct military applicability, e.g., to support foreign or economic policies. When such controls are established, they should use mechanisms other than the MCTL.

Voluntary Controls

Voluntary measures tend both to involve fewer costs than measures founded on regulations and formal sanctions for noncompliance, and to avoid the adversarial atmosphere that accompanies much government regulation. The trial voluntary prepublication review arrangement now in place between the National Security Agency and cryptography researchers is an example--one that some officials see as a possible model for other scientific fields.

Three important characteristics, however, distinguish cryptography from most other fields. First, unlike some other dual-use areas, the military implications of new developments in cryptography are clear to researchers in that field, and thus the danger of open disclosure of particularly critical findings is fully appreciated in the scientific community. Second, the field is small and homogeneous, and the volume of papers produced is limited, so that the government can rely on nonbureaucratic means to interact with researchers. Third, and perhaps most significant, the National Security Agency has an unusually high degree of internal technical competence, which prevents governmental judgments that are needlessly conservative and insensitive to the needs of researchers.

<u>The Panel concludes that the voluntary publication control mechanism developed for cryptography is unlikely to be applicable to other research areas that bear on national security. However, the Panel recommends that consideration be given to adopting this mechanism in future cases if and where the appropriate preconditions exist.</u>

Staffing Deficiencies

Another impediment to the development of workable control measures relates to the adequacy of staffing in government agencies. Budget restrictions have resulted in inadequate numbers of personnel in some cases; reduction-in-force rules may shift inexperienced personnel into positions for which they are not suitably trained; and agencies have had difficulty in attracting people who have the technical background to keep abreast of fast-moving developments in the many relevant scientific and technical fields in their area of responsibility. The shortage of staffing may be particularly significant, for example, in the processing of visa applications.

There are similar difficulties in the intelligence effort that is directed at the leakage problem. The intelligence agencies do not have enough personnel who are able to judge the status of American technical capabilities in specified fields in comparison with those of Soviet bloc nations. Lacking such net assessments, decision makers are sometimes left with no reliable way to evaluate the meaning of a particular technological loss or the value of proposed exchange programs. When in doubt the safe way is to deny an application.

<u>The Panel recommends that, despite the severe budgetary restraints now in effect, serious consideration should be given to increased staffing in situations where it can be demonstrated that an agency's ability to implement, monitor, or enforce regulations, or to give adequate service, is being compromised by lack of a sufficient number of adequately trained people, as, for example, in the case of processing visa applications and developing intelligence assessments.</u>

THE FACTUAL BASIS FOR DECISIONS CAN BE IMPROVED

As would be expected in the case of any newly perceived national policy concern, the dimensions of the current technology transfer problem are not yet fully understood. Nonetheless, decisions of two types are being made--in setting general priorities for the overall national effort and in applying specific controls. There are three particular areas where the factual basis for such decisions needs to be improved in order to avoid wasteful or inadequate policy actions.

As has been seen, there is some concern among government officials that the American research community will become progressively more important as a potential source of leakage in the future. Efforts begun now may help keep U.S. leakage control programs in balance if and when such a shift occurs.

Assessment Capability

As the Panel has sought to gain perspective on the technology transfer problem (particularly with respect to the role of the research community), it has been unable to find adequate information on the nature or extent of the loss of technology. This is in part because federal agencies themselves have not kept adequate data or performed analyses on how frequently controls have been warranted. It is important, for example, to know the number and types of exchange visits that have been disallowed, of instances in which ITAR and EAR have been invoked for research results, and of instances where research projects have unexpectedly developed information that has been classified. The incompleteness of such data denies the government an opportunity to learn quickly about the nature and extent of the U.S. leakage problem, as well as the costs of its control efforts. Better empirical bases for decision making will help the government set program priorities and understand how the benefits to be expected of technology transfer controls are related to their private and governmental costs. The fact that federal efforts to control unwanted technology transfer are dispersed among many independent programs impedes the collection of more data needed. No agency now has a mission responsibility that encourages data collection on the costs of controls.

There is also a need for the generation and central collection of information on the relative strength of the United States and the Soviet bloc in particular scientific fields. Without such assessments it is impossible to evaluate properly the costs and benefits of international scientific cooperation.

The absence of a coordinated program for conducting assessments precludes the orderly review and modification of the nation's control efforts.

<u>The Panel recommends that the government establish a focal point of expertise in basic science and technology for the purpose of evaluating the costs and benefits of scientific openness with respect to existing or proposed restrictions. There is also a need to assess the standing of the United States in comparison with other nations in specific scientific areas. Decisions on visa policy, exchanges, and export restrictions should be based on advice from such assessments. The Office of Science and Technology Policy (OSTP) has the capability to organize this type of effort and is placed sufficiently high in the government science and technology policy hierarchy to make recommendations on such matters.</u>

Technology Transfers to the Third World

The Panel's analysis of the effects of technology transfer on national security has focused on the U.S.-Soviet relationship. This emphasis is in accord with the government's concentration on leakage to the Eastern bloc. The Panel recognizes, however, that similar concerns may develop in U.S. relations with other countries as well. Although it is widely accepted that the participation of U.S. universities in the training of foreign students is desirable because it spreads technology and technical know-how that will improve the economic and social circumstances of less developed countries, some students are from foreign countries that are unfriendly or potentially unfriendly to the United States or its allies. Examples of such countries today include Iran and Libya, but it must be stressed that the list is difficult to compile objectively and is continually changing.

A serious concern is that technology transferred to Third World countries will permit them to develop modern weapons sooner and at lower cost than would otherwise be the case. The most immediate worry in this regard is the proliferation of nuclear weapons. Improved Third World military capabilities--whether nuclear or nonnuclear--may not constitute an immediate threat to U.S. interests, but such developments have destabilized regional balances of power and led to local conflicts in the past, and are likely to do so in the future. The Panel therefore views the potential consequences of North-South technology transfer to be of great significance to future world stability and believes that this issue will continue to pose difficult policy and implementation questions, both for the federal government and for the research community.

Although the Panel did not address the matter of selective treatment of particular Third World countries, it notes that both the current administration and the preceding one have embraced policies indicating less concern about the transfer of dual-use technology, particularly to the People's Republic of China (PRC). The Chinese military capacity to deploy advanced technology is much less than the Soviet capacity, and there is little danger that the PRC will act in concert with Soviet intelligence. In fact, in order to improve relations with the PRC, the government has encouraged attendance at U.S. universities by Chinese students, and there are now approximately 6,000 Chinese students studying here. Despite this government policy, in a number of instances officials in the State Department have asked universities to report on these students' research activities (see Appendix J). The administration of export controls with respect to U.S.-PRC cooperative programs continues to be burdensome. Burdens on the government and on the research community would be greatly reduced if these procedures were further moderated in order to reflect the special status accorded the PRC in official U.S. policy.

For the Third World in general, it may develop that the transfer of militarily significant technologies will prove to have national security implications for the United States that in the long run will match the dangers of leakage to the Soviet bloc. This potential problem is receiving too little attention.

<u>The Panel notes that its deliberations did not extend to the complex issues raised by military-related technology transfer from advanced industrial nations to Third World nations in regionally unstable areas or to those that may be potentially hostile to the United States and its allies. The Panel recommends that this subject receive further attention by the National Academy of Sciences or by other qualified study groups under federal sponsorship.</u>

Review of Scientific Exchange Proposals

As noted above, the most effective potential channel of technology transfer from the research community is the transfer of know-how through long-term working relations and apprenticeships. International exchanges provide an opportunity for such transfers. Review of proposed exchanges of scientific personnel involving American universities is carried out by the Department of State in consultation with the U.S. intelligence community's Committee on Exchanges (COMEX).

There is a need to bring the research community's insight and expertise to this process and to ensure that the research community appreciates the government's reasons for concern about foreign abuse of scientific exchanges.

<u>The Panel recommends that the intelligence and university communities establish an ongoing effort to raise awareness in the scientific community regarding the problems and costs of technological loss, and in the intelligence community regarding the problems and costs of applying restrictions on academic campuses. The Panel recommends the establishment of an academic advisory group to COMEX that would facilitate more effective communication between the universities and the appropriate federal agencies regarding scientific exchanges.</u>

BETTER MUTUAL ACCOMMODATION BETWEEN THE GOVERNMENT AND RESEARCHERS CAN AND MUST BE ACHIEVED

We are still in a period in which the public debate on national security and scientific freedom is noteworthy for its high rhetoric and mutual mistrust. Some officials in the government have contributed to this atmosphere by showing impatience with the traditional openness of research; some people in the research community have as yet failed to acknowledge that the government has just cause for concern with some academic practices.

Government-Science Relations

Some of the disagreement within the United States over the need for more rigorous controls on scientific communication arises from insufficient mutual understanding about the motives, methods of operation, and concerns of the two communities involved. Many people within the U.S. scientific community, for example, have an inadequate understanding of the processes by which technology is transferred to the U.S.S.R. Moreover, many scientists and engineers are largely unaware, except in the most general terms, of the scope of the Soviet intelligence-gathering effort. By the same token, there is evidence that, with important exceptions, some key government officials lack sufficient appreciation of the dynamics that foster scientific progress. In some cases people in government have failed to assess accurately the types of administrative solutions that are feasible on most research campuses.

Both university and industrial researchers have spoken out about the perceived vagueness of existing control mechanisms and the way controls have been implemented. Researchers find themselves suddenly informed that their research results may fall under ITAR and EAR. Because the MCTL is not available to those without secret clearance, many who are subject to ITAR/EAR controls lack crucial information about the specific technologies that are of national security concern to the government. The division of regulatory authority among many agencies and programs has compounded the confusion in the research community.

<u>The Panel recommends that the comprehensive forum proposed originally by the National Commission on Research[2] be brought into existence as soon as practicable, under the auspices of the Academy complex. It further recommends that one of the specific standing responsibilities of the forum be discussion of science and technology</u>

---

[2] National Commission on Research. <u>Accountability: Restoring the Quality of the Partnership.</u> (Washington, D.C.: NRC, March 1980), p. 27.

<u>transfer. The forum should promote exchanges of information and concern among (a) the affected line agencies, namely, the departments of State, Commerce, and Defense, (b) the affected agencies of the intelligence community, (c) the appropriate law enforcement agencies, and (d) representatives of the U.S. scientific community. In the view of the Panel, it is important that the forum meet on a regular basis and that it serve as the basis for the development of less formal and more direct channels of communication and cooperation.</u>

The National Academies' Ad Hoc Committee on Government-University Relationships in Support of Science, which will soon conclude its work, is also expected to recommend that such a forum is needed. This forum is seen as a device that would facilitate improved communication among participants with widely varying goals on a range of key policy issues. It would be an instrument for the prevention of surprise decisions and would facilitate movement toward consensus. The existence of such a forum would not, of course, prevent researchers and/or agency officials from pursuing negotiations on their own, nor would it limit their exercise of the responsibilities of their offices. Its primary purpose would be to achieve mutual understanding of motives, goals, and problems. It would also provide a convenient avenue for the continuous examination and reexamination of issues.

University Involvement in Research Whose Results Will Not Be Disseminated

Universities have attempted to ensure that the results of university research are made available to the scientific community. The Panel has noted, however, a growing interest by some university faculty members and students in research to which access is restricted. Thus far universities seem to have found roles in such research that do not compromise their policies of unrestricted publication. However, universities should ensure that their participation in such research does not evolve in ways that would undermine their principal mission or risk the freedom from which universities derive their strength.[3]

The risk to universities is that the part of the university that

---

[3] The Panel notes that potential restrictions on the dissemination of results are not limited to military research. Many academic fields are of interest to commercial firms who seek to protect a competitive advantage by withholding proprietary information about technology. Many such firms support university research. Current university policies may permit a short delay in the open publication of such research results in order to conduct patent reviews, but generally do not otherwise allow restrictions on access or communication. However, if more severe proprietary restrictions evolve, it must be recognized that they could have the same adverse effects as the national security restrictions discussed in this report.

accepts a role in developing information that cannot be disseminated is, in some respects, no longer an integral part of the university; a figurative wall is erected to keep out those who might use the information for other than its intended purpose. It is in the national interest to prevent universities from transforming themselves into government or commercial research laboratories. There are other organizations that exist for the express purpose of undertaking classified and proprietary commercial research.

<u>The Panel recommends that universities should be vigilant when considering whether to accept research programs that may develop information that is not to be made available to the public, lest they compromise the freedom from which they derive their strength.</u>

U.S.-U.S.S.R. SCIENTIFIC EXCHANGE PROGRAMS SHOULD BE BROUGHT INTO BETTER BALANCE

Scientific exchanges with the Soviet Union can enrich U.S. scientific efforts and help maintain a continuing and reliable assessment of Soviet capabilities and growth. Exchange visits also have significant cultural and intelligence benefits.

There is very little evidence that scientific exchange programs have had an identifiable adverse effect on U.S. national security. This does not imply that the flow of scientific information between the United States and the U.S.S.R. through exchange is balanced. It is not. Some substantial imbalance would be expected, even if there were no concerted Soviet effort to collect information in the United States, given the fact that, on the whole, U.S. science and technology are more advanced than are Soviet science and technology. Still, there is a significant and growing number of scientific fields in which the Soviets have invested heavily and have achieved world-level proficiency.

It should be clearly understood, however, that there are also some significant risks associated with exchange programs. The Soviet Union utilizes exchange programs to collect sensitive information, sometimes with highly specific aims. Heightened sensitivity among U.S. scientists to this fact is desirable.

<u>The U.S. scientific community should recognize the potential for foreign misuse of exchange programs for intelligence purposes. If scientists in academia and elsewhere perceive activities that threaten national security, it is appropriate that they voluntarily inform government officials.</u>

While proposals for particular exchange visits are reviewed with respect to possible technology losses, the design and operation of the exchange programs themselves have been influenced far more by the foreign policy objective. There are ways to improve U.S.-U.S.S.R. exchange programs to better balance their scientific benefits with the risk of potential technology losses.

Bilateral Intergovernmental Agreements

Following recent events in Poland, U.S. policy has been to steadily curtail scientific exchanges under U.S.-U.S.S.R. bilateral agreements. certain fields covered by the bilaterals--for example, plasma physics, condensed matter physics, and fundamental properties of matter--are areas of considerable Soviet strength and provide useful scientific contributions to American research, and can provide insight into Soviet activities. The government should weigh scientific as well as foreign policy factors in decisions about the bilateral agreements.

The Panel urges the administration to become more selective about which programs it chooses to cancel, renew, or allow to expire. These decisions should be made on a substantive basis, and this suggests the need for increased involvement of the White House Office of Science and Technology Policy leading to a more comprehensive examination of the costs and benefits of the exchanges.

Inter-Academy Exchanges (NAS-ASUSSR)

The exchange program between the National Academy of Sciences and the Soviet Academy of Sciences (ASUSSR) has generally been successful. Partly because of the "sending-side-selects" scheme for designating participants, however, there is room for improvement through the participation of a higher proportion of outstanding scientists. Without such improvement, the risk of technology loss through abuses of the program might eventually equal the value of the exchanges to this nation. Movement in the direction of "receiving side selects" on both sides would improve the program.

The Panel recommends that at least 50 percent of the visitors on both sides should be invited by the receiving side, with invitations based on publications and other measures of competence of the visitors. Agreements should contain a clause that would allow cancellation of the program if it is determined that the other side is not sending those agreed upon or abuses the exchange program for intelligence purposes.

International Research and Exchange Board Program

Most U.S. participants have found their International Research and Exchange Board (IREX) program visits to be satisfactory, and the program clearly advances U.S. graduate training in Soviet studies. A majority of the U.S. hosts have reported Soviet students' scientific performance to be at least satisfactory. Nonetheless, the commonly perceived asymmetry and the possible abuse of the program for intelligence gathering purposes are noted.

The Panel recommends that (a) some fixed portion of the IREX program be reserved for technical and scientific fields in which the

United States and the U.S.S.R. have rough parity; (b) review procedures on the receiving side be enhanced to ensure that only bona fide scholars are sent on the exchanges; (c) all militarily sensitive areas be excluded from the exchanges by formal agreement; and (d) new or expanded procedures be developed to ensure that the program is mutually beneficial.

6

COMPILATION OF RECOMMENDATIONS

Chapters 4 and 5 set out the Panel's recommendations and the reasoning behind them. This chapter provides a compilation of our principal recommendations.

CONTROL OF UNIVERSITY RESEARCH ACTIVITIES

Unrestricted Areas of Research

The Panel recommends that no restrictions of any kind limiting access or communication should be applied to any area of university research, be it basic or applied, unless it involves a technology meeting all of the following criteria:

- The technology is developing rapidly, and the time from basic science to application is short;
- The technology has identifiable direct military applications; or it is dual-use and involves process or production-related techniques;
- Transfer of the technology would give the U.S.S.R. a significant near-term military advantage; and
- The United States is the only source of information about the technology, or other friendly nations that could also be the source have control systems as secure as ours.

Classification

The Panel recommends that if government-supported research demonstrably will lead to military products in a short time, classifications should be considered. It should be noted that most universities will not undertake classified work, and some will undertake it only in off-campus facilities.

Gray Areas

The Panel recommends that in the limited number of instances in which all of the above four criteria are met but in which classification is unwarranted, the values of open science can be preserved and the needs of government can be met by written agreements no more restrictive than the following:

a. Prohibition of direct participation in government-supported research projects by nationals of designated foreign countries, with no attempt made to limit physical access to university space or facilities or enrollment in any classroom course of study. Moreover, where such prohibition has been imposed by visa or contractually agreed upon, it is not inappropriate for government-university contracts to permit the government to ask a university to report those instances coming to the university's attention in which the stipulated foreign nationals seek participation in any such activities, however supported. It is recognized that some universities will regard such reporting requests as objectionable. Such requests, however, should not require surveillance or monitoring of foreign nationals by the universities.

b. Submission of stipulated manuscripts simultaneously to the publisher and to the federal agency contract officer, with the federal agency then having 60 days to seek modifications in the manuscript. The review period is not intended to give the government the power to order changes: The right and freedom to publish remain with the university, as they do with all unclassified research. This does not, of course, detract from the government's ultimate power to classify in accordance with law any research it has supported.

The Panel recommends that in cases where the government places such restrictions on scientific communication through contracts or other written agreements, it should be obligated to record and tabulate the instances of those restrictions on a regular basis.

The provisions of EAR and ITAR should not be invoked to deal with gray areas in government-funded university research.

THE WORKABILITY OF EXPORT CONTROLS ON SCIENTIFIC COMMUNICATION

Export of Domestically Available Technical Data under ITAR and EAR

The Panel recommends that unclassified information that is available domestically should receive a general license (exemption) from the formal licensing process.

Scope of ITAR and EAR Technical Data Provisions

The Panel recommends that information that is not directly and significantly connected with technology critical to national security should also receive a general license (exemption) from the formal licensing process. The critical technology list approach--if carefully formulated--could serve to define those limited areas where controls are appropriate.

The MCTL

The Panel recommends a drastic streamlining of the MCTL by reducing its overall size to concentrate on technologies that are truly critical to national security. The Panel recommends that items should be removed from the MCTL if they are in one or more of the following categories:

    a. Science and technology whose transfer would not lead to a significant near-term improvement in Soviet defense capability;

    b. Science underlying a mature technology--that is, a technology that is evolving slowly;

    c. Science underlying dual-use technology that is not process-oriented;

    d. Components used in militarily sensitive devices that in themselves are not sensitive.

The Panel recognizes that technology transfer controls may be adopted for reasons other than direct military applicability, e.g., to support foreign or economic policies. When such controls are established, they should use mechanisms other than the MCTL.

Voluntary Controls

The Panel concludes that the voluntary publication control mechanism developed for cryptography is unlikely to be applicable to other research areas that bear on national security. However, the Panel recommends that consideration be given to adopting this mechanism in future cases, if and where the appropriate preconditions exist.

Staffing

The Panel recommends that, despite the severe budgetary restraints now in effect, serious consideration should be given to increased staffing in situations where it can be demonstrated that an agency's ability to implement, monitor, or enforce regulations, or to give adequate service, is being compromised by lack of a sufficient number of adequately trained people, as for example, in the case of processing visa applications and developing intelligence assessments.

DATA FOR DECISION MAKING

Assessment Capability

The Panel recommends that the government establish a focal point of expertise in basic science and technology for the purpose of evaluating the costs and benefits of scientific openness with respect to existing or proposed restrictions. There is also a need to assess the standing of the United States in comparison with other nations in specific scientific areas. Decisions on visa policy, exchanges, and export restrictions should be based on advice from such assessments. The Office of Science and Technology Policy (OSTP) has the capability to organize this type of effort and is placed sufficiently high in the government science and technology policy hierarchy to make recommendations on such matters.

Technology Transfers to the Third World

The Panel notes that its deliberations did not extend to the complex issues raised by military-related technology transfer from advanced industrial nations to Third World nations in regionally unstable areas or to those that may be potentially hostile to the United States and its allies. The Panel recommends that this subject receive further attention by the National Academy of Sciences or by other qualified study groups under federal sponsorship.

Review of Scientific Exchange Proposals

The Panel recommends that the intelligence and university communities establish an ongoing effort to raise awareness in the scientific community regarding the problems and costs of technological loss, and in the intelligence community regarding the problems and costs of applying restrictions on academic campuses. The Panel recommends the establishment of an academic advisory group to COMEX that would facilitate more effective communication between the universities and the appropriate federal agencies regarding scientific exchanges.

THE GOVERNMENT-UNIVERSITY RELATIONSHIP

Government-University Forum

The Panel recommends that the comprehensive forum proposed originally by the National Commission on Research be brought into existence as soon as practicable, under the auspices of the Academy complex. It further recommends that one of the specific standing responsibilities of the forum be discussion of science and technology transfer. The forum should promote exchanges of information and concern among (a) the affected line agencies, namely, the departments of State, Commerce, and

Defense, (b) the affected agencies of the intelligence community, (c) the appropriate law enforcement agencies, and (d) representatives of the U.S. scientific community. In the view of the Panel, it is important that the forum meet on a regular basis and that it serve as the basis for the development of less formal and more direct channels of communication and cooperation.

University Involvement in Research Whose Results Will Not Be Disseminated

The Panel recommends that universities should be vigilant when considering whether to accept research programs that may develop information that is not to be made available to the public, lest they compromise the freedom from which they derive their strength.

U.S.-U.S.S.R. SCIENTIFIC EXCHANGES

Heightened Awareness

The U.S. scientific community should recognize the potential for foreign misuse of exchange programs for intelligence purposes. If scientists in academia or elsewhere become aware of activities that threaten national security, it is appropriate that they voluntarily inform government officials.

Bilateral Intergovernmental Agreements

The Panel urges the administration to become more selective about which programs it chooses to cancel, renew, or allow to expire. These decisions should be made on a substantive basis, and this suggests the need for increased involvement of the White House Office of Science and Technology Policy leading to a more comprehensive examination of the costs and benefits of the exchanges.

Inter-Academy Exchanges (NAS-ASUSSR)

The Panel recommends that at least 50 percent of the visitors on both sides of inter-Academy exchanges should be invited by the receiving side, with invitations based on publications and other measures of competence of the visitors. Agreements should contain a clause that would allow cancellation of the program if it is determined that the other side is not sending those agreed upon or abuses the exchange program for intelligence purposes.

International Research and Exchange Board Program

The Panel recommends that (a) some fixed portion of the IREX program be reserved for technical and scientific fields in which the United States and the U.S.S.R. have rough parity; (b) review procedures on the receiving side be enhanced to ensure that only bona fide scholars are sent on the exchanges; (c) all militarily sensitive areas be excluded from the exchanges by formal agreement; and (d) new or expanded procedures be developed to ensure that the program is mutually beneficial.

ADDITIONAL COMMENT BY
HAROLD T. SHAPIRO

While fully concurring with the recommendations of our Panel's report, I would note that the report takes as a given the overall strategic parameters of current U.S. defense policy, including, for example, general attitudes toward security classification as a means of constraining the flow of science or technology. In the context of the complex set of issues raised both by the nuclear realities of our time and changing economic and scientific relations among nations, this proposition is not self-evident. Analysis of these issues, however, was beyond the scope of this report.

LIST OF BRIEFERS, CONTRIBUTORS, AND LIAISON REPRESENTATIVES

Briefers

ARTHUR J. ALEXANDER, Associate Head, Economics Department, The Rand Corporation
BETSY ANDERSON, Consular Officer, Bureau of Consular Affairs, Department of State
LEWIS M. BRANSCOMB, Chief Scientist, IBM Corporation
STEPHEN D. BRYEN, Deputy Assistant Secretary, International Economic, Trade and Security Policy, Department of Defense
WILLIAM D. CAREY, Executive Officer, American Association for the Advancement of Science
MICHAEL CIFRINO, Attorney Advisor, Office of the Assistant General Counsel, Department of Defense
W. DONALD COOKE, Vice President for Research, Cornell University
JOHN C. CROWLEY, Director, Federal Relations for Science and Research, Association of American Universities
JAMES DEARLOVE, Chairman, Committee on Exchanges, Technology Transfer Branch; Defense Intelligence Agency, Department of Defense
BOHDAN DENYSYK, Deputy Assistant Secretary, Export Administration, Department of Commerce
ERWIN FRIEDLANDER, Staff Physicist, Lawrence Berkeley Laboratory
ALBERT GORE, JR., Chairman, Investigations and Oversight Subcommittee, Committee on Science and Technology, House of Representatives
WALTER GRANT, Chief, Technology Transfer Branch of the Nuclear Energy and Applied Science Division, Defense Intelligence Agency, Department of Defense
C. DAVID HARTMANN, Executive Secretary, Technology Transfer Intelligence Committee
MARTIN HELLMAN, Professor, Department of Electrical Engineering, Stanford University
CHARLES HORNER, Deputy Assistant Secretary for Science and Technology, Bureau of Oceans and International Environmental and Scientific Affairs, Department of State
BOBBY RAY INMAN, Deputy Director, Central Intelligence Agency
ERNEST B. JOHNSTON, Senior Deputy Assistant Secretary, Bureau of Economic and Business Affairs, Department of State

FRANCIS B. KAPPER, Director, Military Technology Sharing, International Programs and Technology, Office of the Under Secretary of Defense for Research and Engineering, Department of Defense
MICHAEL LORENZO, Deputy Under Secretary of Defense for Research and Engineering (International Programs and Technology), Department of Defense
MICHAEL B. MARKS, Special Assistant to the Under Secretary, Office of the Under Secretary for Security Assistance, Science and Technology, Department of State
RICHARD F. POST, Deputy Associate Director for Physics, Magnetic Fusion Division, Lawrence Livermore National Laboratory
FRANK H. T. RHODES, President, Cornell University
RONALD RIVEST, Professor, Department of Electrical Engineering and Computer Science, Massachusetts Institute of Technology
HOWARD E. ROSENBLUM, Deputy Director for Communications Security, National Security Agency
JOSEPH P. SMALDONE, Chief, Arms Licensing Division, Office of Munitions Contol, Department of State
RICHARD SPICER, Intelligence Analyst, Soviet Section, Intelligence Division, Federal Bureau of Investigation
STEPHEN UNGER, Professor, Department of Computer Science, Columbia University
JACK VORONA, Assistant Vice Director for Scientific and Technical Intelligence (International), Defense Intelligence Agency, Department of Defense
DAVID A. WILSON, President's Executive Assistant, University of California
LEO YOUNG, Director for Research and Technical Information, Office of the Under Secretary of Defense for Research and Engineering, Department of Defense

Contributors

LAURENCE J. ADAMS, Senior Vice President, Martin Marietta Corporation
WAYNE BERT, Munitions Policy Analyst, International Economic, Trade and Security Policy, Department of Defense
JENNIFER SUE BOND, Program Analyst, National Science Foundation
J. FRED BUCY, President, Texas Instruments, Inc.
ALAN M. CAMPBELL, Executive Secretary, U.S.-U.S.S.R. Committee on Cooperation in Physics, Office of International Affairs, National Academy of Sciences
ROSEMARY CHALK, Program Head for Scientific Freedom and Responsibility, American Association for the Advancement of Science
JOHN C. CROWLEY, Director, Federal Relations for Science and Research, Association of American Universities
EDWARD E. DAVID, JR., President, Exxon Research and Engineering Company
CAROLE A. GANZ, International Science Analyst, National Science Foundation
RICHARD L. GARWIN, IBM Fellow, T. J. Watson Research Center, IBM Corporation

S. E. GOODMAN, Professor, Department of Information Systems and
    Decision Sciences, University of Arizona
RUTH GREENSTEIN, Associate General Counsel, Policy, National Science
    Foundation
WILLIAM C. HITTINGER, Executive Vice President, RCA Corporation
JEANNE E. HUDSON, Special Assistant, Office of the Director, National
    Science Foundation
JOHN W. KISER III, Kiser Research, Inc.
RICHARD KRASNOW, Congressional Science Fellow
LAWRENCE C. MITCHELL, Staff Director, Office of International Affairs,
    National Academy of Sciences
MARTIN E. PACKARD, Assistant to the Board Chairman, Varian Associates
THOMAS O. PAINE, Thomas Paine Associates
HAROLD RELYEA, Analyst, Government Division, Congressional Research
    Service
LEONARD M. RIESER, Chairperson, Committee on Scientific Freedom and
    Responsibility, American Association for the Advancement of Science
IAN M. ROSS, President, Bell Laboratories
ROBERT D. SCHMIDT, Vice Chairman of the Board, Control Data Corporation
ROLAND W. SCHMITT, Vice President, Corporate Research and Development,
    General Electric Company
MICHAEL A. STROSCIO, Special Assistant to the Director of Research and
    Technical Information, Office of the Under Secretary of Defense for
    Research and Engineering, Department of Defense

Liaison Representatives

    American Academy of Arts and Sciences

        HERMAN FESHBACH, Professor, Department of Physics, Massachusetts
            Institute of Technology

    American Association for the Advancement of Science

        J. THOMAS RATCHFORD, Associate Executive Officer, American
            Association for the Advancement of Science

    American Chemical Society

        RAYMOND P. MARIELLA, Executive Director, American Chemical Society

    American Geophysical Union

        FRED SPILHAUS, Executive Director, American Geophysical Union

    American Physical Society

        MELVIN B. GOTTLIEB, Science and Public Policy Fellow, The Brookings
            Institution

Association of American Universities

THOMAS A. BARTLETT, President, Association of American Universities

Association of American Universities-Department of Defense Forum

DAVID A. WILSON, President's Executive Assistant, University of California

Department of Commerce

BOHDAN DENYSYK, Deputy Assistant Secretary, Export Administration, Department of Commerce

Department of Defense

STEPHEN D. BRYEN, Deputy Assistant Secretary, International Economic, Trade and Security Policy, Department of Defense
FRANCIS B. KAPPER, Director, Military Technology Sharing, International Programs and Technology, Office of the Under Secretary of Defense for Research and Engineering, Department of Defense
LEO YOUNG, Director for Research and Technical Information, Office of the Under Secretary of Defense for Research and Engineering, Department of Defense

Department of State

MICHAEL B. MARKS, Special Assistant to the Under Secretary, Office of the Under Secretary for Security Assistance, Science and Technology, Department of State

Intelligence Community

JAN P. HERRING, Chairman, Technology Transfer Intelligence Committee

National Aeronautics and Space Administration

BURTON I. EDELSON, Associate Administrator for Space Science and Applications, NASA
JACK KERREBROCK, Associate Administrator for Office of Aeronautics and Space Technology, NASA

National Science Foundation

DONALD N. LANGENBERG, Deputy Director, National Science Foundation

Office of Science and Technology Policy

EDWARD MCGAFFIGAN, Assistant Director for International Affairs, Office of Science and Technology Policy

The Institute of Electrical and Electronics Engineers, Inc.

ROBERT P. BRISKMAN, Assistant Vice President, COMSAT General Corporation

LIST OF ACRONYMS

AAAS	American Association for the Advancement of Science
AAU	American Association of Universities
ASUSSR	Academy of Sciences of the U.S.S.R.
COCOM	Coordinating Committee for national export controls
COMEX	Committee on Exchanges
DARPA	Defense Advanced Research Projects Agency
EAA	Export Administration Act
EAR	Export Administration Regulations
IREX	International Research and Exchange Board
ITAR	International Traffic in Arms Regulations
MCTL	Militarily Critical Technologies List
NAS	National Academy of Sciences
OEA	Office of Export Administration
OMC	Office of Munitions Control
PCSG	Public Cryptography Study Group
PRC	People's Republic of China
TTIC	Technology Transfer Intelligence Committee

ANNOTATED BIBLIOGRAPHY

CONTENTS

1. Perspectives on the Current Problem

    1.1 Background on East-West science and technology transfer
    1.2 Recent correspondence and speeches
    1.3 Government documents

2. Mechanisms for Controlling Technology Transfer

    2.1 Classification
    2.2 Export controls
    2.3 Contractual controls in government-funded research
    2.4 Voluntary restraints
    2.5 Visa control

3. Information Restrictions and U.S. Scientific and Technological Enterprise

4. Legal Issues

5. Case Examples

    5.1 Very high speed integrated circuits (VHSIC)
    5.2 Cryptography
    5.3 Other scientific and technological areas

6. Working Papers of the Panel

1. PERSPECTIVES ON THE CURRENT PROBLEM

1.1 Background on East-West science and technology transfer

Bucy, J. Fred. "Technology Transfer and East-West Trade: A Reappraisal," International Security 5:3 (Winter 1980/81), 132-151. Examines the political, economic, and security dimensions of the U.S. failure to develop a coherent and sound export control policy. Considers also the impact on relations with other Western allies.

Gustafson, Thane. Rand Report. Selling the Russians the Rope? Soviet Technology Policy and U.S. Export Controls. Santa Monica, CA: The Rand Corporation, April 1981. 77 pp. Aims to take a new and critical look at the objectives and assumptions of U.S. high-technology export-control policy. Describes the main developments in Soviet technology policy over the last 10 years, analyzes the reasons for Soviet technology lag, and draws the implications for U.S. policy.

Gustafson, Thane. "Why Doesn't Soviet Science Do Better Than It Does?" in Linda L. Lubrano and Susan G. Solomon, eds., The Social Context of Soviet Science (Boulder, Colorado: Westview Press, 1980), pp. 31-68. Analyzes the organization, financing, and training of personnel for Soviet science and its attendant strengths and weaknesses.

Kiser, John W. III. Commercial Technology Transfer from Eastern Europe to the United States and Western Europe. Report prepared for the Department of State, Washington, D.C.: Kiser Research, 1980. 115 pp. Considers commercial technology transfer from four East European countries—Czechoslovakia, Hungary, the German Democratic Republic, and Poland—to the United States and selected Western European countries. The report focuses on benefits derived by the United States, principally from buying licences.

Kiser, John W. III. Report on the Potential for Technology Transfer from the Soviet Union to the United States. Report prepared for the Department of State and the National Science Foundation, Washington, D.C.: Department of State, 1977. 168 pp. Examines the premise that the Soviet Union has scientific information and technology available and of use to the West. Establishes a historical record on the sale of Soviet technology.

Kiser, John W. III. "Tapping Eastern Bloc Technology," Harvard Business Review 60:2 (March-April 1982), 85-93. Suggests the notion of reverse technology transfer from East to West. Points out some of the difficulties of obtaining technology from the Communist world.

National Research Council, Board on International Scientific Exchanges, Commission on International Relations. Review of U.S.-U.S.S.R. Inter-Academy Exchanges and Relations, September 1977. Report of a 2-year study conducted under the chairmanship of Carl Kaysen. Various sections treat the development of the program, the description of its features, and the status of exchanges in several fields of science. Conclusions are included as to the costs and benefits of scientific exchanges.

National Research Council, Commission on International Relations. <u>Summary Report on the Special Meeting on NAS Relations with the Soviet and East European Scientific Communities and Academies of Sciences</u>, October 28, 1981. 4 pp. This summary report is based on discussions held at NAS at an ad hoc meeting called at the invitation of President Frank Press and chaired by Herbert F. York. The meeting included 22 participants from universities, industry, journalism, and foundations. It considered the impact on East-West scientific interactions of the changed international (and domestic) political and economic environments.

National Research Council, Board on International Scientific Exchanges, Commission on International Relations. <u>Review of the U.S./U.S.S.R. Agreement on Cooperation in the Fields of Science and Technology</u>, May 1977. Evaluates the effectiveness of the Academy-managed exchange agreement with the Soviet Union and makes specific recommendations regarding the terms and arrangements for its continuation.

Relyea, Harold C. "Business, Trade Secrets, and Information Access Policy Developments in Other Countries: An Overview," <u>Adminstrative Law Review</u> 34:2 (Spring 1982), 315-371. Presents capsule description of existing or emerging policy concerning the right of access to official information or records held by governments in Western Europe, Australia, New Zealand, Canada, and Scandanavia. Special consideration is given to the implications for business, commercial records, and trade data. A final section explores the issue of transborder data flows.

Relyea, Harold C. National Security Controls and Scientific Information. Congressional Issue Brief Number IB82083, updated August 18, 1982. 15 pp. Succinct general policy background paper. Bibliography.

Skolnikoff, Eugene B. "Technology Transfer to Other Countries: Life-Threatening or Unimportant?" unpublished, April 22, 1982. This essay by the director of the Center for International Studies at M.I.T., examines the overall question of technology transfers to other countries and assesses the costs and benefits of more stringent control measures.

U.S. Congress, Office of Technology Assessment. <u>Technology and East-West Trade</u>, Chapter IX, "The East-West Trade Policies of America's COCOM Allies." Washington, D.C.: GPO, Nov. 1979. pp. 173-202. Comparative discussion of the trade policies adhered to by the principal Western allies of the United States: West Germany, France, United Kingdom, and Japan. A general background document.

Zaleski, Eugène, and Helgard Wienert. <u>Technology Transfer between East and West</u>. Paris: OECD, 1980. 435 pp. [Washington, D.C: sold by OECD Publications and Information Center.] Topics include: historical perspectives on East-West trade, statistical

analyses, forms of technology transfer, Eastern and Western
policies regarding technology transfer, influence of transfers
on Eastern economies, effect of economic factors on transfer,
and effect of transfers on Western economies.

1.2 Recent correspondence and speeches

Brady, Lawrence J. "Taking Back the Rope, Technology Transfer and U.S. Security." Statement before the Association of Former Intelligence Officers, March 29, 1982.

Carey, William D. "Scientific Exchanges and U.S. National Security." Science 215 (January 8, 1982), 139-141. Letters exchanged by Mr. Carey, Executive Officer and Publisher of Science, and The Honorable Frank Carlucci, Deputy Secretary of the Department of Defense, concerning scientific exchanges, conferences, and the unclassified, open scientific literature.

Carey, William D. "Science and the National Security." Science 214 (November 6, 1981), 609. Comment on perceived concerns of military officials toward technology transfer.

"High Tech Censorship." Transcript of the 1982 MacNeil-Lehrer Report, Corporation for Public Broadcasting, April 21, 1982. Interview and debate with George Davida (cryptographer), Daniel Schwartz (National Security Counsel), Stephen Bryen (DOD), and William Carey (AAAS).

Inman, Admiral B. R., and William D. Carey. "Classifying Science: A Government Proposal . . . And a Scientist's Objection," Aviation Week and Space Technology 116:6 (February 8, 1982), 10-11. Inman's address to the Annual Meeting of the American Association for the Advancement of Science, Washington, D.C., January 7, 1982; rebutted by Executive Officer of AAAS on subject of science and secrecy.

Inman, Admiral B. R. "National Security and Technical Information." Speech to the Annual Meeting of the American Association for the Advancement of Science, Washington, D.C., January 7, 1982. 7 pp. Presented morning session, "Striking a Balance: Scientific Freedom and National Security."

Kennedy, Donald. Letter on behalf of the Presidents of Cornell University, M.I.T., Cal Tech, the University of California, and Stanford University to Secretaries Malcolm Baldridge, Alexander Haig, and Caspar Weinberger, February 27, 1981.

Press, Frank. Statement before the Subcommittee on Science, Research and Technology and Subcommittee on Investigations and Oversight of the U.S. House of Representatives, Committee on Science and Technology, March 29, 1982.

Weinberger, Caspar W. "Technology Transfers to the Soviet Union," The Wall Street Journal, January 12, 1982. p. 32. Support and explanation of administration views on need for export controls.

1.3 Government documents

1.3.1 U.S. Congress

Hart, Gary W., U.S. Senator (D-Col). "High Technology Trade Act of 1982," S. 2356, Congressional Record (Senate, April 1, 1982). 7 pp. Description of a proposed bill that offers a different view concerning how the United States can maintain its technological edge.

U.S. Congress, Senate, Committee on Banking, Subcommittee on International Finance and Monetary Policy. "Hearing on Export Controls for National Security Purposes," April 14, 1982. Testimony of Lawrence Brady (Commerce), Fred C. Ikle (DOD), Ernest B. Johnston (State), and Edward J. O'Malley (FBI).

1.3.2 Central Intelligence Agency

U.S. Central Intelligence Agency. "Soviet Acquisition of Western Technology," April 1982. 15 pp. Describes the Soviet program to acquire U.S. and Western technology, the acquisition mechanisms used, the spectrum of Western technology that has contributed to Soviet military capability, and the problems of restricting the transfer of Western technological information.

1.3.3 Department of Defense

Lorenzo, Michael, Deputy Under Secretary of Defense for Research and Engineering. "The Role and Responsibilities of Defense Research and Engineering in Export Control." Statement before the Permanent Subcommittee on Investigations at hearings on Transfer of United States, High Technology to the Soviet Union and Soviet Bloc Nations. 97th Congress, 2nd session. Washington, D.C.: GPO, 1982. Committee on Governmental Affairs, U.S. Senate, May 11, 1982.

Perle, Richard N. "The Soviet Connection," Defense 82 (February 1982), 10-15. (Adapted from congressional testimony of November 12, 1981.) Provides a limited (i.e. unclassified) statement of the Defense Department's concerns about Soviet access to U.S. science and technology.

U.S. Department of Defense, Office of the Director of Defense Research and Engineering. Report of the Defense Science Board Task Force on Export of U.S. Technology, An Analysis of Export Control of U.S. Technology--A DOD Perspective. Washington, D.C.: GPO, 1976. 39 pp. This is the so-called Bucy report, which examines a number of critical technologies, their impact on U.S. strategic requirements, the mechanisms through which information about them is transferred, and the current effectiveness of export controls and the COCOM agreement.

U.S. Department of Defense. Soviet Military Power. Washington, D.C.: GPO. 99 pp. Chapter VI, "Quest for Technological Superiority," is a key source for the current administration's arguments for controlling transfer of technology.

2. MECHANISMS FOR CONTROLLING TECHNOLOGY TRANSFER

2.1 Classification

Ehlke, Richard C., and Harold C. Relyea. The Freedom of Information Improvements Act of 1981--Proposed Amendments of the Reagan Administration: A Brief Analysis and Commentary. Washington, D.C.: Congressional Research Service, Library of Congress, January 22, 1982. 81 pp.

Executive Office of the President. Executive Order No. 12065, "National Security Information." Federal Register 43: 128 (June 28, 1978), 28949-28961.

Halperin, Morton H., and Allan Adler. Comment on Draft Executive Order on National Security Information--and related material. February 9, 1982

Office of the Press Secretary, White House. "Executive Order on National Security Information." April 2, 1982

U.S. Department of Defense, Office of the Director of Defense Research and Engineering. "Report of the Defense Science Board, Task Force on Secrecy," July 1, 1970. 12 pp. This is the "Seitz Report," which considered the matter of classification from several viewpoints but focused primarily on classification of scientific and technical information. It assessed the positive and negative aspects of classification, the types of information that need to be classified, and the length of time classification should be maintained.

U.S. House of Representatives Committee on Government Operations. Executive Order on Security Classifiation. Washington, D.C.: USCIPO, 1982. 363 pp. Hearing before the English Subcommittee on Government Information and Individual Rights, March 10 and May 5, 1982.

U.S. House of Representatives, Committee on Government Operations. "The Government's Classification of Private Ideas." 96th Congress, 2d Session, December 22, 1980. 244 pp. U.S. House Committee on Science and Technology testimony concerning the impact of national security considerations on science and technology. Witnesses included Admiral B. R. Inman, Lawrence J. Brady, George Millburn, Frank Press, Robert Corell, Edward Gerjuoy, and John McLucas.

2.2 Export controls

Conahan, Frank C., Director, International Division, G.A.O. Statement Before the Subcommittee on International Finance and Monetary Policy, Senate Committee on Banking, Housing and Urban Affairs, "The Administration of Export Controls under the Export Administration Act," April 30, 1981. Hearings on International Affairs Functions of the Treasury and the Export Administration Act. 97th Congress, 1st session. Washington D.C.: GPO, 1982. Provides a critical analysis of the administration of export controls, including the constraints imposed by the necessity to seek compromise within COCOM and the inefficiencies of the bureaucratic review process.

Eagle Research Group, Inc. Report of the United States Munitions List Study, Prepared for the Office of Munitions Control, Department of State, ERG 81-123F1, April 14, 1981. This study was conducted for the State Department for the purpose of providing an analytical input into the final report to be submitted to the Congress on the MCTL as required by the International Security and Development Cooperation Act of 1980.

Gustafson, Thane. "U.S. Export Controls and Soviet Technology," Technology Review, 85:2 (February/March 1982), 34-35. Examines whether the critical technologies approach can improve the export control system.

International Trade Administration, U.S. Department of Commerce. "Overview of the Export Administration Program." October 1981. 17 pp. Provides a short summary of the legislative history, administrative organization, and enforcement procedures relating to the Export Administration Regulations. Also deals with interagency consultation and cooperation.

Legislative History, Export Administration Act of 1981, P.L. 97-145. 12 pp.

Office of the Secretary of Defense. "Initial Militarily Critical Technologies List," Federal Register 45: 192 (October 1, 1980), 65014-65019.

Packard, Martin E., "A Businessman's View of the Effect of Export Licensing on Technology Transfer to the USSR," unpublished, 1981. Examines the many sources of technological information and the effectiveness of various control measures. Considers the costs and benefits of export licensing.

U.S. Congress, Senate Committee on Governmental Affairs. Transfer of United States Technology to the Soviet Union and Soviet Bloc Nations. Statement of Fred Asselin, Staff Investigator, Permanent Subcommittee on Investigations, at hearings held by the subcommittee, May 4, 5, 6, 11 and 12, 1982. 97th Congress, 2nd Session. Washington, D.C.: GPO, 1982. 655 pp. This is a report of an investigation conducted by the staff of the Permanent Investigations Subcommittee on the effectiveness of the Department of Commerce in enforcing the Export Administration Regulations. It is highly critical of current enforcement practices.

U.S. Department of State "Proposed Revision of the International International Traffic on Arms Regulations," *Federal Register*, 45: 246 (December 19, 1980), 83970-83995.

U.S. General Accounting Office. *Export Controls: Need to Clarify Policy and Simplify Administration*. ID-79-16. Washington, D.C.: GPO, March 1, 1979. 67 pp. This report examines the decision-making apparatus for determining what technology or products must be controlled and the effectiveness of this system. It assesses both domestic and multilateral export control policies, and includes an analysis of COCOM control procedures.

U.S. General Accounting Office. *Export Control Regulation Could Be Reduced Without Affecting National Security*. ID-82-14. Washington, D.C.: GPO, May 26, 1982. This report examines the process of review for export applications and considers ways in which the process could be streamlined without affecting U.S. national security. The report also discusses inefficiencies in the licensing review process and government efforts to curtail illegal export activity.

U.S. General Accounting Office. *U.S. Munitions Export Controls Need Improvement*. ID-78-62. Washington, D.C.: GPO, April 25, 1979. 48 pp. + 7 appendixes. This report recommends ways to improve munitions export controls and to provide assurance that such exports conform to law and authorized munitions export licenses. It examines the workload and licensing procedures employed by the Office of Munitions Control in the State Department.

2.3 Contractual controls in government-funded research

U.S. Department of Defense, Office of the Under Secretary of Defense for Research and Engineering. "Report of the Defense Science Board Task Force on University Responsiveness." January 1982. 14 pp. Prepared at the request of the U.S. House of Representatives, Committee on Armed Services. Includes evaluation of the impact of implementation of contractual controls on research dissemination.

2.4 Voluntary restraints

Berry, R. L. "Academic Freedom and Peer Reviews of Research Proposals and Papers," *American Journal of Agricultural Economics* 62:4 (November 1980), 639-646. Advantages and disadvantages of peer reviews, including administrative use for suppression of unpopular proposals and papers as constituting censorship unless justifiable. Bibliography.

Kolata, Gina Bari. "Prior Restraints Recommended," *Science* 211 (February 20, 1981), 797. Public Cryptography Study Group's proposal for voluntary system of prior restraints.

2.5 Visa control

"New Pressure on Scientific Exchanges," Science 215 (February 5, 1982), 637-638. Recent administration actions affecting exchanges.

3. INFORMATION RESTRICTIONS AND U.S. SCIENTIFIC AND TECHNOLOGICAL ENTERPRISE

AAAS Committee on Scientific Freedom and Responsibility National Security and Scientific Communication. Memo, June 1982. 8 pp. A summary of responses received by the Committee to letters from 100 leading American scientists and engineers on the topic of science and secrecy.

Branscomb, Lewis M. Letter to Leonard M. Rieser, Chairman, AAAS Committee on Scientific Freedom and Responsibility, April 5, 1982. 3 pp. Presents a personal view of the conflict between national security and unclassified research with particular reference to impacts on the university and business communities.

Center for Science and Technology Policy, Graduate School of Business Administration, New York University. Current Issues in Export Controls of Technology, Background Information and Summary of Discussion, November 1981. 38 pp. Considers several issues involving the use of export controls to restrict the flow of technology. Presents the results of faculty discussions as to the most critical questions and impacts on university/industrial research program.

Denning, Peter J. "A Scientist's View of Government Control Over Scientific Publication." Paper presented at the annual meeting of the American Association for the Advancement of Science, Washington, D.C., January 7, 1982.

Gray, Paul E. "Technology Transfer at Issue: The Academic Viewpoint," IEEE Spectrum (May 1982), 64-68. Delineates the arguments presented to the Departments of State, Commerce, and Defense by the group of five university presidents.

National Science Foundation. Foreign Participation in U.S. Science and Engineering Higher Education and Labor Markets, Special Report NSF 81-316. September 1981. Examines the evidence regarding the growth of foreign participation in U.S. science and engineering graduate school programs. Concentrates on graduate training, doctorate production, and postdoctorates.

Nelkin, Dorothy. "Intellectual Property: The Control of Scientific Information," Science 216 (May 14, 1982), 704-708. A review of diverse situations that have led to disputes and of efforts to establish principles for contolling intellectual property.

Unger, Stephen H. "The Growing Threat of Government Secrecy," Technology Review (February/March 1982). Background summary paper on expansion of barriers being erected to the free flow of scientific information. Brief Bibliography.

U.S. House of Representatives, Committee on Science and
Technology. *Impact of National Security Considerations on
Science and Technology*. Washington, D.C.: GPO, 1982. U.S.
House, Committee on Government Operations annotation: Based on
a study made by the Subcommittee on Government Information and
Individual Rights. Address the issue of invention secrecy,
public cryptography, and atomic energy restricted data.

Wallich, Paul. "Technology Transfer at Issue: The Industry
Viewpoint," *IEEE Spectrum* (May 1982), 69-73. Identifies the
nature of the commercial technology export problem and the
position of the private sector.

4. LEGAL ISSUES

Cheh, Mary M. "Government Control of Private Ideas." Paper
presented at the Annual Meeting of the American Association for
the Advancement of Science, Washington, D.C., January 8, 1982.

Green, Harold P. "Where the Balance Has Been Struck--Information
Control Under the Atomic Energy Act." Paper presented at the
Annual Meeting of the American Association for the Advancement
of Science, Washington, D.C., January 7, 1982.

Greenstein, Ruth, "National Security Controls on Scientific
Information," unpublished, 1982. Analyzes the use of export
controls to restrict free exchange of scientific information,
particularly that which is only indirectly related to controlled
hardware. Addresses the question of whether an export control
system can be designed that meets national security objectives
while maintaining a vital scientific base.

Olson, Theodore B., Office of the Assistant Attorney General.
Memorandum on Export Administration Regulations for Henry D.
Mitman, Director, Capital Goods Production Materials Divisions,
Department of Commerce. July 28, 1981. 6 pp.

Olson, Theodore B., Office of the Assistant Attorney General.
Memorandum on Constitutionality of the Proposed Revision of the
Technical Data Provisions of the International Traffic in Arms
Regulations for William B. Robinson, Office of Munitions
Control, Department of State. July 1, 1981. 16 pp.

5. CASE EXAMPLES

5.1 Very high speed integrated circuits (VHSIC)

"Controls Sought on Technology Exports," *Aviation Week and Space
Technology* 114:7 (February 16, 1981), 85. Defense Department
steps to prevent transfer of technology in VHSIC program.

Martin, Jim, "Very High Speed Integrated Circuits--Into the Second
Generation, Part 1: The Birth of the Program," *Military
Electronics/ Countermeasures* 7: 12 (December 1981), 52-58, 71-73.

Martin, Jim, "Very High Speed Integrated Circuits--Into the Second Generation, Part 2: Entering Phase 1," <u>Military Electronics/Countermeasures</u> 8:1 (January 1982), 60-66.

Sumney, Larry W. Memorandum for VHSIC Program Directors, December 12, 1980. 2 pp. Interim guidance from the Director of DOD VHSIC Program Office concerning the applicability of ITAR and EAR to VHSIC research.

Vanderheiden, Robert M. "VHSIC: Midterm Report on a Dynamic Circuit Program," <u>Defense Electronics</u> 14: 2 (February 1982), 54-62.

5.2 Cryptography

Kahn, David. "Cryptology Goes Public," <u>Foreign Affairs</u> 58:1 (Fall, 1979), 141-159. Detailed overview of national security and private sector conflict over development of cryptology. NSA activities, communications security, eavesdropping, countermeasures, government regulatory activity, DES, secrecy orders, issues.

Kahn, David. "The Public's Secrets," <u>Progressive</u> 44:11 (November 1980), 27-31. Spread of cryptology and concerns of U.S. intelligence agencies regarding national security. Patent secrecy orders, export controls and scientific meetings. Overview of issues.

Schwartz, Daniel C. "Scientific Freedom and National Security--A Case Study of Cryptography." Paper presented at the annual meeting of the American Association for the Advancement of Science, Washington, D.C., January 7, 1982.

5.3 Other scientific and technological areas

Channon, Stanley L. <u>Status and Recommendations for Export Control of Composite Materials Technology</u>, IDA Paper P-1592. 2 Vols. Institute for Defense Analysis, Science and Technology Division, September 1981. 519 pp. This report presents the results of a 27-month study of U.S. and foreign technology in organic matrix, metal-matrix, carbon-carbon, and ceramic-matrix composite materials and a critical review of the relevant U.S. export control regulations. The advantages and disadvantages of export control and the effects of these controls on industrial innovation, academic research, and international technical communications are discussed. Suggested methods for handling proprietary information, emerging technology, and the involvement of foreign nationals in advanced composite materials technology are presented.

Goodman, S. E. Memorandum on U.S. computer export control policies: value conflicts and policy choice. 51 pp. Reviews U.S. export controls for computer products and know-how and examines the policy choices.

"Of Bubbles, Bombs, and Batteries: Secrecy Snafus," <u>Technology Review</u> 85:2. (February/March 1982), 36-39, 84-85. Review of recent export actions, including The Progressive case, conferences, and secrecy orders.

"Pajaro Dunes Biotechnology Statement," Tech Talk (M.I.T.) 26:31 (April 7, 1982), 8. Preliminary consensus position resulting from the Conference on Biotechnology, Pajaro Dunes, California, March 25-27, 1982. The conference was attended by high-level representatives from Stanford, Cal Tech, University of California, Harvard, and M.I.T.

6. WORKING PAPERS OF THE PANEL

[Photocopies of the collected working papers of the Panel on Scientific Communication and National Security are available from the National Academy Press, 2101 Constitution Avenue, N.W., Washington, D.C. 20418.]

Alexander, Arthur J. "Soviet Science and Weapons Acquisition."

Cooke, W. D., Thomas Eisner, Thomas Everhart, Franklin A. Long, Benjamin Widom, and Edward Wolf. "Restrictions on Academic Research and the National Interest."

Kiser, John W. III. "East-West Technology Transfer."

Post, Richard F., Melvin B. Gottlieb, and Wolfang K. H. Panofsky. "Comments on Historical Aspects of Classification and Communication in Magnetic Fusion Research."

Wallerstein, Mitchel B. "The Office of Strategic Information (OSI), U.S. Department of Commerce, 1954-1957."

Wallerstein, Mitchel B. "The Coordinating Committee for National Export Controls (COCOM)," with Annex by John P. Hardt and Kate S. Tomlinson.

Appendix A

MEMORANDUM FROM THE INTELLIGENCE SUBPANEL
TO THE PANEL ON SCIENTIFIC COMMUNICATION AND NATIONAL SECURITY
(UNCLASSIFIED VERSION)

J. Deutch, J. Killian, F. Lindsay,
W. K. Panofsky, S. Phillips, E. Staats

The full Panel is charged to examine the question "What is the effect on national security of technology transfer to adversary nations by means of open scientific communications, either through scientific literature or by person-to-person communications. . . ?" In effect the subpanel is to query "What has been the effect on national security of technology transfer. . . ?" The subpanel has held two meetings with members of the intelligence community (on May 5, 1982, and June 21, 1982). The subpanel was not concerned with the effect of free scientific exchange on U.S. technology, scientific progress, and international goodwill; the only positive effect considered was that on U.S. intelligence. The subpanel's function was to gather information on the assigned topic attainable at a security level higher than that accessible by the full Panel. The question of the effect on national security of either greater or lesser restraints than those now practiced, a matter of concern to the full Panel, was not examined by the subpanel.

SUMMARY FINDING

While there has been a serious transfer of U.S. technology to the Soviet Union from many sources that is directly relevant to military systems, there is a strong consensus that the universities, as well as open communication involving the university community, appear to be a very small part of this problem up to the present time. At the same time open information on basic research, which is an essential part of

---

Appendix A is the unclassified version of a memorandum to the full Panel from the subpanel, which is made up of six members of the Panel who hold clearance at the highest level. Although the original memorandum is classified, it is available at the National Academy of Sciences to those who hold appropriate security clearance.

our open society and research process, has without doubt contributed to the scientific base of the Soviet Union as well as other nations.

This lack of concrete evidence linking the academic community and other scientific communication channels to specific losses of militarily relevant technology does not imply a lack of clear Soviet intent to use such open scientific communication channels to increase their military potential. The marginal evidence on our subject is submerged by the security losses through outright espionage targeted on U.S. systems, in particular in foreign countries, by outright illegal conduct by individuals or corporations in international trade, and by secondary transfers of actual material from legal or illegal recipients abroad to adversary destinations. Intellience efforts have not specifically focused on the "open communication" component of the technology transfer problem, but have given highest priority to localizing the other, larger, channels of technology loss.

The exception is foreign visitors from Communist countries where the intelligence community, through the interagency Committee on Exchange (COMEX), has been active for many years. Person-to-person communication involving U.S. researchers is one of many channels for the transfer of sensitive technologies, and, compared to other transfer mechanisms, the potential loss there of sensitive technologies has been limited. In part this is because there have been for many years U.S. government mechanisms that try to assess the likely technology transfer balance of proposed exchange programs. When U.S. government monies are involved, the government has received and reviewed proposed research programs well before foreign Communist visitors have been due to arrive. When necessary, programs have been denied. In many more instances, the program have been modified in some way to lessen access to sensitive technologies.

Such efforts to limit access also have occurred regarding foreign Communist students whose financing did not involve U.S. government monies, but in these instances the effectiveness of proposed limitations relied even more on the cooperation of the academic hosts. Since the mid-1950s, COMEX has provided such information, analysis, and advice to the Department of State and other government agencies regarding technology transfer and other implications of the proposed programs of foreign Communist students and other visitors.

The fact that few demonstrable losses of direct military significance from U.S. academic and other free exchange sources have been detected in the past does not, of course, prove that more significant losses will not occur in the future; in particular, should university activities extend further into areas of direct military applicability. The intelligence community believes there is a clear trend toward greater Soviet bloc effort in acquiring basic technology associated with universities. Thus the problem of technology transfer from universities is dynamic; this may lead to greater Soviet emphasis on acquisition of technology with long-term applications in the future. However, this subpanel has seen little evidence that the issue which the full Panel is charged to address has been important in the past in the total context of the loss of military technology.

SOVIET EFFORTS

Acquisition of Western technology has been a goal of the Soviet state throughout its history and has been a facet of the Russian tradition before the revolution. However, the effort to collect foreign technology, including basic science and technology, has become highly organized and targeted in recent years. This effort is directed from the highest level of government, in particuar through the powerful Military-Industrial Commission (VPK), which is the coordinating agency for all military R&D, and the State Committee for Science and Technology (GKNT).

The Soviets and Eastern bloc deploy intelligence officers to many countries, including the United States, to collect scientific and technical information. Individual students and scholars nominated to participate in exchange programs in the West are most often screened by their respective intelligence services. Additionally, Third World students are often questioned by Soviet intelligence for open information and may be recruited for intelligence purposes.

Techniques for collecting science and technology information are both overt and covert. The Soviets gather whatever they can from open sources, and then target that which remains for illegal purchase or the use of classic espionage. One senior intelligence community official has publicly suggested that the Soviets and East European intelligence services have been involved in the acquisition of about 70 percent of the militarily useful, militarily related technologies that have been acquired from the West. They have used clandestine, technical, and overt collection techniques in the process. Of the remaining 20 to 30 percent of the acquisition of information of potential direct military value to the Soviets, most comes through legal purchases and open-source publications acquired by other Soviet organizations. The same offical advises that a very small percentage of such military technology is acquired from direct technical exchanges conducted by scientists and students. This subpanel agrees with that observation.

Since the late 1970s, there has been an increased emphasis on the acquisition of new Western technologies emerging from universities and other research centers. The Soviets presumably also make full use of access to advanced technologies provided by various exchange arrangements with Western European countries and with Japan.

A small percent of the several thousand Soviets entering the United States annually under some sort of exchange arrangement are known to have some intelligence affiliation. The number of individuals having intelligence tasks among the scientific exchanges is significantly higher. For example, a substantial number of these Soviet personnel have been identified participating in the International Research and Exchanges Board Graduate Student and Young Faculty program with various U.S. universities.

The Soviet Union devotes an enormous effort--perhaps involving 100,000 people--to sifting and systematically disseminating unclassified technical materials from the West and Japan, such as those available from the National Technical Information Service (NTIS) in the United States.

EXAMPLES OF SOVIET COLLECTION FROM U.S. ACADEMIC SOURCES

Specific evidence of Soviet collection of technology information from U.S. academic and other free exchange sources relates almost exclusively to episodes of abuses by Soviet or Soviet bloc visitors of their guest status in the United States. These abuses are in a number of categories which might be tabulated in the order of severity as follows:

1. The visitor's technical activities and studies go beyond the agreed field of study.
2. The visitor's time during the period of study is poorly accounted for, or excessive time is spent in library activities collecting information not related to the agreed fields of study.
3. The visitor, either successfully or unsuccessfully, attempts to evade the restrictions imposed on the program itinerary.
4. The visitor participates in clearly illegal activities, such as intelligence "drops"; attempts to examine secure containers; etc.

The intelligence community remains concerned about Communist acquisition of sensitive technology even when nothing illegal outside the agreed upon study program is likely to take place.

The recognized episodes of abuse by Soviet visitors of their guest status have been disturbing but have not led to evidence of significant consequences. Inadequate handling of visitors by U.S. government or academic authorities has at times contributed to the potential for abuse. From examination of the episodes it is difficult to secure evidence that any significant losses of U.S. critical technology have occurred to the benefit of identifiable Soviet military systems. Part of this lack of evidence is, of course, due to the fact that all specifically traced technology losses of military importance, some of which have indeed been very serious, have occurred in a nonuniversity and nonfree communication context. Therefore, trying to identify the losses of relevance to the charge to the subpanel is a "needle in the haystack" problem. Moreover it must be acknowledged that tracing technology loss to a specific item of military hardware would be a most difficult matter.

EFFECTIVENESS OF SOVIET ASSIMILATION OF DATA FROM U.S. ACADEMIC AND OTHER "FREE EXCHANGE" SOURCES

While the subpanel has not examined this issue in detail, it has seen no meaningful intelligence data on the effectiveness of the internal transfer process to direct the data flow from the massive U.S.S.R. foreign technology collection effort into military application. The subpanel is skeptical that the organized Soviet effort involving tens of thousands of people charged with digesting the vast volume of open literature is an effective means to expedite technology transfer.

The broader question as to the ability of the Soviet military R&D process to absorb new technology--be it generated at home or acquired abroad--was not examined by the subpanel.

The lack of evidence of an identifiable significant detrimental effect to U.S. security from international scientific communication does not mean, of course, that the net information flow in the science and technology areas resulting from these exchanges between the United States and the Soviet Union and other Warsaw Pact countries is balanced. It is not. However, the imbalance is not different from what one would expect from the fact that on the whole, United States performance in most relevant fields of science is higher than that of the Soviet Union. The net information flow is that expected from the existence of openly available information in basic scientific areas and is not directly traceable to targeted collection activities. The imbalance in information flow is more significant in connection with exchanges involving personnel of lower skills. In contrast, the general consensus is that information exchange and intelligence collection is considerably more balanced in exchanges involving more advanced professional personnel. There the prebriefing and special selection of Soviet exchange visitors may provide some Soviet advantage.

The intelligence community also cautions that those exchanges removed from the umbrella of official interacademy and intergovernmental exchanges would be considerably more difficult to monitor. There has been significant intelligence reporting of Soviet military technology from U.S. participants in scientific exchange programs.

Privately sponsored exchange programs are increasing. The federal government has little leverage on these exchanges. Visas, up to the present, have not been denied on the basis of projected technology loss. Enforcement of travel restrictions on Soviets is difficult, and movement of East Europeans is not controlled by imposed travel restrictions.

Only a relatively small portion of the exchange arrangements screened by U.S. government mechanisms are judged to be of significant concern because of the potential for unwanted technology transfer. For example, COMEX has conducted formal reviews of the programs of only about 7 percent of the programs it screens, judging that the balance would not constitute a significant problem. Of those formally reviewed, about one-third are judged to pose significant technology transfer problems, and perhaps one-half are judged to offer "some" concerns. However, most often suggestions are offered by the government merely to modify the proposed program or itinerary to lessen the technology transfer concerns, and the exchanges proceed.

COMMUNICATION BETWEEN THE INTELLIGENCE COMMUNITY AND THE RESEARCH COMMUNITY

The subpanel is impressed by the fact that some members of the intelligence community involved in assessing the technology transfer problem have little acquaintance with the workings of the research community or the conduct of basic research. We hasten to add that there are notable and important counterexamples to this statement. The subpanel also notes that many members of the academic community have little appreciation of the constructive and necessary role of the intelligence community in assessing foreign activities.

This "communication gap" manifests itself in several respects. The subpanel finds that some members of the intelligence community interpret such activities as excessive use of the library or lack of total dedication by a Soviet visitor to his projected task to be suspicious conduct. By such criteria most American researchers would seem suspect at their own research campuses. Conversely, the subpanel observes that some members of the U.S. research community are at times totally insensitive to national security issues and uncooperative with representatives of U.S. intelligence agencies. Reports on visitor activities or on visits by U.S. scientists travelling abroad are frequently late and at times not made at all, even if required by government contract.

Appendix B

THE HISTORICAL CONTEXT OF NATIONAL SECURITY CONCERNS ABOUT SCIENCE AND TECHNOLOGY

Mitchel B. Wallerstein
Staff Consultant

Prior to World War II, only the armed services and the departments of War and State maintained security classification programs. These were designed to protect military secrets and to safeguard diplomatic communications. The legal authority for these programs was derived from a general administrative statute.

In September 1942, however, the Office of War Information issued a government-wide regulation on creating and administering classified materials. The principal responsibility for military research during the war was assigned to the Office of Scientific Research and Development (OSRD), directed by Vannevar Bush. Basically, OSRD adopted the security classification system used by the armed services. However, because of the rigid requirements with respect to handling, transmitting, and filing data, OSRD tried to avoid top secret assignments, particularly within the university environment. Most OSRD projects were, in fact, classified at lower levels, such as confidential or secret.

When the war ended, a problem arose with respect to the declassification and release of scientific and industrial data obtained in Germany and Japan by allied forces. President Truman decided that these spoils of war should be released promptly, but that in doing so close attention should be paid to national security, given growing Soviet belligerence.

A somewhat parallel situation also arose when OSRD faced the problem of publishing the large mass of information that had accumulated during the five years of wartime scientific silence. The resulting OSRD "summary technical reports" were very broad in the scope of topics covered and consequently were placed under tight security restrictions. Only 250 copies of the reports ultimately were printed. But OSRD did approve public disclosure of some of the results of wartime research,

---

The material in this appendix is drawn from information provided to the Panel during its briefing meetings and a variety of primary and secondary source documents. It has not been possible, however, to undertake the kind of exhaustive, historical review necessary to ascertain the detailed accuracy of every event cited. Rather, the intent here is to convey to the reader a sense of the order and flow of the major developments that have contributed the present situation.

such as that undertaken at the M.I.T. Radiation Laboratory, which was the primary U.S. microwave radar research facility. The publication of these materials proved to be of enormous benefit to universities and industry, both in the United States and abroad (including the Soviet Union), in the further development of microwave and related technology.

SECURITY AND TRADE RESTRAINTS IN THE POSTWAR PERIOD

In the period after World War II, as the ideological struggle between the United States and the Soviet Union intensified, the federal government became increasingly concerned about protecting scientific information. The Atomic Energy Act of 1946, for example, precluded public dissemination of most of the results of the Manhattan District Project or subsequent atomic research. The act, which was amended in 1954, included a "born secret" provision, meaning that all information about atomic energy was automatically classified at the moment of its creation. In 1950 President Truman issued an executive order that contained a vaguely defined standard for protecting national security as the rationale for classifying secret documents. This justification on the grounds of a need to protect national security has continued to the present day, although the definition of national security has been modified several times.

Government restraints on the movement of goods out of the United States also originated during World War II. In fact, export controls have existed in one form or another since July 1940. Although it originally had been anticipated that these restrictions would be terminated at the end of the war, the advent of the Cold War prompted the passage of the Export Control Act of 1949. This act, which remained in effect for the next 20 years, provided for continuing examination of exports to the Soviet Union and most other Communist countries. The 1949 act was succeeded by the Export Administration Act of 1969, and more recently by the Export Administration Act of 1979. All three laws have been implemented by the Department of Commerce through a comprehensive set of procedures known as the Export Administration Regulations (EAR), which are in turn used to administer the Commodity Control List, an extensive itemization of restricted products and processes. Although modified by subsequent acts (see Chapter 3, Volume I of this report), the original Export Control Act required the government to prevent the exportation of goods that might assist either the economic or military potential of communist countries.

Another method of controlling the export of security-related goods was developed in 1954, when the International Traffic in Arms Regulations (ITAR) were established, initially as part of the Mutual Security Act of 1954. Administered by the Department of State, the ITAR rules are used to control the export of military systems, including the "design, production, manufacture, repair, overhaul, processing, engineering, development, operation, maintenance or reconstruction of . . . implements of war on the U.S. Munitions List" or "any technology that advances the state of the art or establishes a new art in any area of significant military applicability." The current foundation for the ITAR is the Arms Export Control Act of 1976.

In order to control the movement of militarily sensitive goods at the international level, the Coordinating Committee for national export controls (COCOM) was established by informal agreement in 1949 (see the annex to this appendix). COCOM, which comprises all the NATO countries except Iceland, plus Japan, has provided a forum for the consideration of trade controls on exports to the Warsaw Pact countries and the People's Republic of China. COCOM is a voluntary organization, and its decisions can only be implemented through the national policies of its members. These national policies sometimes differ significantly. COCOM maintains three separate lists covering munitions, atomic energy, and dual-use items. The latter accounts for a majority of the trade matters considered by the group.

For a brief period during the 1950s there was a third dimension to the U.S. effort to regulate the flow of information and goods to the Soviet Union and other potential adversary nations. At the behest of the National Security Council, the Office of Strategic Information (OSI) was established within the Department of Commerce in 1954 (see Panel Working Paper on OSI, available from the National Academy Press). Although never authorized by legislation, OSI was created by the Eisenhower administration because of growing concern about Soviet efforts to obtain U.S. industrial and military information. Ultimately, however, OSI ran afoul of both the Department of Defense, which viewed its security role as redundant, and Congress, which was concerned about OSI's negative impact on scientific projects. As a result, OSI ceased operations in June 1957.

SECRECY AND LOYALTY DURING THE 1950s

In addition to the efforts to protect sensitive technological information through classification and through export restrictions, other approaches have also been taken to prevent the transfer of information. The Internal Security Act of 1950 (the so-called McCarran Act) and the Immigration and Naturalization Act of 1952 (the so-called McCarran-Walter Act) established rigid and indiscriminant restrictions on the issuance of visas to aliens seeking to enter the United States. One result of these two laws was that large numbers of distinguished European scientists found it much more difficult to visit the United States to attend meetings or to assume appointments at American universities. In some cases visas were refused outright; in others visas were approved only after such long delays that the scientific meeting had already taken place or the offer of a teaching appointment had been withdrawn.

At the same time, Congress became concerned about the loyalty of scientists conducting unclassified research with the aid of federal grants, primarily from the National Science Foundation (NSF) and the National Institutes of Health (NIH). As a result, Congress placed increasing pressure on both NSF and NIH to adopt restrictive policies, particularly in the form of required loyalty oaths for those receiving grants. NIH apparently did draft an oath-signing procedure (which was never implemented), but the National Science Board at NSF rejected the idea and reaffirmed its support of the principle that approval of grants should be based on the "experience, competence, and integrity" of those seeking grants, "based on the judgments of scientists having a working knowledge" of an applicant's qualifications. Ultimately, the question of the need for loyalty oaths was referred to the National Academy of Sciences (NAS) at the request of presidential assistant Sherman Adams. An NAS committee, under the chairmanship of J. A. Stratton of M.I.T., subsequently recommended against the use of special loyalty requirements for persons conducting unclassified scientific research. The NAS committee also proposed specific criteria for defining government policy on loyalty matters in scientific research. These recommendations were accepted by the Eisenhower administration and were made part of executive branch policy in 1956.

Two other efforts to protect scientific secrets also deserve mention here. The first was the passage in 1951 of the Patent and Invention Secrecy Act. Under the terms of this legislation, which still remains in effect, the Patent Office is charged with sending all requests for patents that may have military applications to the Department of Defense. The Patent Office is empowered to block the granting of such a patent and to prohibit the inventor from disclosing the invention to anyone else. In FY 1979, about 5 percent of the over 100,000 patent applications apparently were sent to defense agencies for review. These reviews resulted in 243 secrecy orders, approximately 40 of which pertained to unclassified research and development. In addition approximately 3,300 secrecy orders were renewed in FY 1979.[1]

The second secrecy-related effort was an executive order issued by President Eisenhower on the classification of secret documents. Like subsequent directives issued by Presidents Nixon, Carter, and Reagan, the Eisenhower order was intended to adjust the classification system to the needs of the current administration. In the case of the Eisenhower order this actually meant a certain degree of relaxation of the classification system used during the Truman period.

THE IMPACT OF DETENTE ON SCIENTIFIC COMMUNICATION

The 1957 launching of Sputnik by the Soviet Union stimulated an enormous increase in the federal government's investment in scientific research, training, and facilities. This support was further strengthened during the Kennedy administration, which held that a robust scientific enterprise was critical to the maintenance of national prosperity and national security. In addition, President Kennedy made the race for the moon the centerpiece of his administration's science and technology effort.

While science and technology were being emphasized for their importance to continued security and prosperity, some rudimentary but nevertheless significant initiatives were being undertaken to expand

---

[1] Stephen H. Unger, "The Growing Threat of Government Secrecy," *Technology Review* (February/March 1982), p. 37.

scientific cooperation between the United States and the Soviet Union. Chief among these was the decision of both sides in 1958 to declassify certain aspects of research on nuclear fusion and to share the results publicly. This action is generally credited with significantly advancing the state of the art in fusion research, as well as with establishing a useful precedent for future scientific exchanges.

Approximately one year later, in July 1959, scientific exchanges between the United States and the Soviet Union were formalized in a historic agreement between the NAS and the Academy of Science of the U.S.S.R. (ASUSSR). The agreement provided for exchange visits by scientists of both nations and for joint symposia, of which eleven were held between 1961 and 1979. Despite this modest thaw in relations, however, it was not until the early 1970s that the United States was able to rid itself fully of its Sputnik era fears, having demonstrated by that time its clear technological superiority through the lunar landing and other achievements. Thus, cooperation in science and technology became an increasingly attractive instrument of foreign policy in the evolving detente with the Soviet Union.

A clear indication of this changed environment was a new U.S.-U.S.S.R. agreement on interacademy cooperation in 1972. Between 1972 and 1974 eleven bilateral intergovernmental agreements in science and technology were also concluded between the two countries, marking a major increase in the level of contact. This shift in attitude was further reinforced by the report of the Task Force on Secrecy of the Defense Science Board, chaired by a former NAS president, Dr. Frederick Seitz. The task force recommended a significant modification of U.S. policy on the classification of secret materials, including a significant decrease in the amount of information classified and in the length of the restrictions.

The keystone of detente, however, was in the realm of trade. A fundamental tenet of the foreign economic policy of the Nixon administration, a policy continued under Presidents Ford and Carter, was the belief that Soviet adventurism could be constrained through an explicit policy of linkage whereby U.S. trade with the U.S.S.R. would be expanded in return for tacit Soviet agreement to abide by the status quo in international affairs. What clearly interested the Soviets beyond all else was greater access to developments in the emerging high-technology industries. As a result, the Export Administration Act of 1969 openly encouraged trade with all nations, including communist countries. The result was a substantial increase in U.S.-Soviet trade, much of it involving dual-use technology, such as computer hardware and ball-bearing grinder machinery.

GROWING CONCERN ABOUT TECHNOLOGY LOSS

By the mid-1970s, however, some disturbing new trends had begun to emerge, both with respect to the configuration of U.S. scientific and technological enterprise and with regard to the U.S.-Soviet trade relationship. In many fields at the cutting edge of science, the distinction between basic and applied research was becoming less relevant. In microelectronics, for example, fundamental physics research increasingly was being carried on side-by-side with the development of industrial applications of that research, which became known as production "recipes." Furthermore, an increasing number of technologies were dual-use in character, and in many cases it was difficult, if not impossible, to separate military applications from civilian ones. And there were growing indications that the Soviet Union, through both legal and illegal channels, was making special efforts to acquire information about dual-use technologies and, wherever possible, to obtain access to production know-how, i.e., to the recipes.

In view of these developments the Defense Science Board commissioned a task force chaired by J. Fred Bucy, president of Texas Instruments Corporation, to examine the entire question of controlling exports of U.S. technology. The task force report, <u>An Analysis of Export Control of U.S. Technology--A DOD Perspective</u>, called for a break with past practices. Basically, the Bucy task force argued that, with the exception of technologies of direct military value to potential adversaries, efforts to control exports should not focus on the <u>products</u> of technology but on <u>design</u> and <u>manufacturing</u> know-how. The report recommended that primary emphasis should be placed on (1) arrays of design and manufacturing know-how; (2) keystone manufacturing, inspection, and test equipment; and (3) products requiring sophisticated operation, application, or maintenance know-how. The Bucy task force concluded that preservation of the U.S. lead in critical technological areas was becoming increasingly difficult but that it could be maintained--first, by denying exportation for technology when it represented a revolutionary (rather than evolutionary) advance for the receiving nation, and second, by strengthening existing export control laws and the COCOM agreement.

Within the federal government, bureaucratic and legislative efforts were undertaken to gain better control over the movement of scientific and technological information out of the United States. The conception of restricted items was broadened to include not only products on the U.S. Munitions List but also technical data relating to those items. As defined under ITAR, "technical data" include the shipping, mailing or carrying by hand of various types of data outside the United States, the disclosure of such data by American citizens visiting abroad, or the disclosure of such data to foreign nationals in the United States during plant visits, briefings, or symposia. Enactment of the Arms Control Act of 1976 and the Nuclear Nonproliferation Act of 1978 both were intended, in part, to impose restrictions on the movement of goods and information related to militarily critical technologies outside of the United States.

It was recognized during the Carter administration that the U.S. intelligence community was poorly equipped to make informed judgments about the potential costs and benefits of granting visas to scientific visitors from the Soviet Union and other East European counties, many of whom were known to have been tasked to acquire scientific and technological information. Accordingly, in 1981 the Technology Transfer Intelligence Committee (TTIC) was established, which incorporated the Committee on Exchanges (COMEX) as well as other relevant agencies of the intelligence community. COMEX itself, which advises the Department of State about the acceptability of foreign individuals proposed under exchange programs, had been in existence since the mid-1950s. COMEX meetings are attended by representatives of all the various agencies within the intelligence community as well as the line agencies having a responsibility or interest in the matter.

Perhaps the most significant recent government initiative was the passage of the Export Administration Act of 1979, which was intended to change the focus of the U.S. export system from an emphasis on goods to an emphasis on technologies. That part of the 1979 legislation that most clearly reflects the new perspectives articulated in the Bucy task force report was the provision calling for the creation of a "Militarily Critical Technologies List" (MCTL). The purpose of the list is to identify technological elements essential to an advanced military capability, with emphasis on manufacturing know-how, keystone manufacturing equipment, goods which contain sophisticated technology, and maintenance know-how. A classified document of many hundreds of pages, the MCTL was developed by the Department of Defense with input from other line agencies and is intended to serve as a guide for modification of the EAR Commodity Control List as well as the lists maintained under the COCOM agreements.

SCIENCE AND TECHNOLOGY COMMUNICATION IN THE POSTDETENTE ERA

By the last year of the Carter administration the East-West political climate had deteriorated substantially because of the Soviet invasion of Afghanistan, the resulting American grain and technology embargoes, and the internal exile of Soviet physicist Andrei Sakharov. These events cast a chill over scientific communications between the United States and the Soviet Union. Moreover, beginning early in 1980, the federal government began to upgrade its efforts to control the dissemination of unclassified scientific and technical information to foreign nationals. These efforts began in February with the International Conference on Bubble Memory. The sponsor of the conference, the American Vacuum Society, was informed that it would have to obtain an export license before admitting Communist bloc scientists to the meeting. That same month a much larger meeting on inertial-confinement fusion research, sponsored by the Optical Society of America and the Institute of Electrical and Electronics Engineers, also became the subject of government control efforts. Conference organizers were informed by the State Department that eight Soviet scientists would be denied visas to attend the meeting.

Heightened government sensitivity to the technology transfer issue also led to increasing difficulties in U.S.-Soviet scientific exchange programs. In mid-December 1981 the National Academy of Sciences notified Stanford, Wisconsin, Ohio State, and Auburn universities that the State Department had requested certain restrictions on a visit to the campuses by Dr. N. V. Umnov, an expert on robotics. These restrictions would have given Umnov access to work in robotics only at the theoretical level, prevented him from visiting industrial facilities, and denied him access to production research or any classified or unclassified research funded by DOD. Stanford replied that it would be unable to receive Dr. Umnov under these restrictions, and researchers at all four universities indicated that the restrictions conflicted with the spirit of open scientific communication. It ultimately proved impossible to work out a compromise acceptable to both the universities and the government. The Umnov case is representative of a number of similar problems that have arisen in recent years (see Appendix J).

Some steps have been taken to develop a modus vivendi between government concern about security and the interest of the universities in open scientific communication. The problem was recently addressed, for example, by another task force of the Defense Science Board. The Task Force on University Responsiveness was chaired by Dr. Ivan Bennett of the New York University School of Medicine, and it considered the new and serious problems now facing the nation's universities in the current economic climate and assessed the impact of these constraints on the capacity of universities to undertake DOD-funded research. The Bennett task force also took up directly the question of applying export controls to academic activities and the effectiveness of this mechanism in limiting the loss of sensitive information.

Another example of an attempt to reconcile the interests of the universities and the federal government came about as a result of concern expressed by the National Security Agency that the publication of certain information on new encryption methodologies in the field of cryptography might violate the Arms Export Control Act of 1976, since cryptography is classified as a munition (see Appendix E). After several years of often difficult debate between government officials and university researchers, the Public Cryptography Study Group (PCSG) was created by the American Council on Education and funded by the National Science Foundation. Under the arrangement subsequently recommended by the PCSG a trial system was established whereby the National Security Agency invites authors to submit papers voluntarily for prior review. A cryptographer who disagrees with the agency's views on the paper can appeal to a standing committee composed of two members appointed by NSA and three appointed by the President's science advisor.

In other fields of research, however, the conflicting views of the government and university researchers have proven more difficult to resolve. In December 1980, for example, the director of the Very High Speed Integrated Circuit (VHSIC) program of DOD released a memorandum raising the possibility that ITAR and EAR might be invoked to regulate the release of unclassified technical research data. The memorandum also suggested that only U.S. citizens and immigrant aliens should be permitted access to DOD-supported research projects. This memorandum brought a protest by the presidents of Stanford University, the California Institute of Technology, M.I.T., Cornell University, and the University of California, who said that:

> It should be recognized that the only realistic way to "contain" VHSIC research is to classify the whole program. In our view

this would be a self-defeating effort: the science underlying
high technologies cannot be put back into the bottle. Furthermore, most universities have concluded that performance of
classified research is incompatible with their essential
purposes. University scientists would pefer, for the most
part, to change their field of interest rather than have their
research and teaching so constrained. [The full text of this
letter is included in Section II, Appendix G.]

As a result of the protest, a Defense Science Board task force on
VHSIC was convened under the chairmanship of William Perry, former
Under Secretary of Defense for Research and Engineering. This task
force recently made the following recommendations to DOD on the
application of export controls to VHSIC research: (1) no controls on
basic research, (2) research with commercial proprietary value should
be subject to EAR, (3) dual-use research that has distinct military
sensitivity should be regulated under ITAR, and (4) single-use defense
technology should be classified.

Similar conflicts between the government and the scientific community occurred later in 1981. The Department of Defense released a
report, <u>Soviet Military Power</u>, which was highly critical of the
technology transfer occurring as a result of scientific exchange
programs, international conferences and symposia, unclassified research
reports, and publication of articles in scientific journals. William
D. Carey, Executive Officer of the American Association for the
Advancement of Science (AAAS), subsequently argued that the same
characteristics of open scientific communication criticized in the
report contributed to the superiority of American technology and hence
to U.S. military strength. Carey's statement was published in the
January 8, 1982, issue of <u>Science</u>, which also included a response by
Frank Carlucci, Deputy Scretary of Defense, who elaborated on DOD's
position. Carlucci contended that the Soviet Union sends scientists to
the United States who are "often directly involved in applied military
research."

In an article published in the <u>Wall Street Journal</u> on January 12,
1982, Secretary of Defense Caspar Weinberger stated that the Soviets
"have organized a massive, systematic effort to get advanced technology
from the West. The purpose is to support the Soviet military buildup."
During that same month, Admiral B. R. Inman, then Deputy Director of
the Central Intelligence Agency, suggested to the annual meeting of the
AAAS that the scientific community should be more cooperative in
voluntarily submitting research results to prepublication review by
appropriate government agencies because of the threat posed by advances
in Soviet science and technology.

One of the results of this debate was the decision in February 1982
to establish a panel under the auspices of the NAS to make an objective
and balanced assessment of the evidence. While this study has been
under way, however, the government has taken two initiatives to restrain
the flow of information. It has proposed amending the Freedom of Information Act to exempt certain categories of information from disclosure,
and it has issued a new executive order on security classification that

frees the government from the obligation to show due cause when it makes classification decisions, thereby reversing the policy of the Carter administration.

RECENT DEVELOPMENTS

A substantial number of other developments pertaining to technology transfer have occurred during the last year and a half:

* In the wake of the Soviet-supported imposition of martial law in Poland, President Reagan ordered the cancellation of some U.S.-U.S.S.R. bilateral scientific exchange programs and the nonrenewal of others. Interacademy scientific exchanges between the NAS and the ASUSSR have also been curtailed due to the displeasure of U.S. scientists over the violation of the human rights of Andrei Sakharov.
* The President has also ordered, as an economic sanction, that all validated export licenses for the Soviet Union be suspended, including exports of U.S. gas pipeline technology.
* At the Ottawa Economic Summit in July 1981 President Reagan asked for greater cooperation among the COCOM allies in restricting technological flows to the Eastern bloc. This resulted in the first high-level meeting of COCOM in over 20 years. (The matter of COCOM is elaborated in the Working Papers of the Panel, which are available from the National Academy Press.)
* In April 1982 the Institute for Scientific Information (ISI), was informed by the U.S. Customs Service that one of the weekly tapes of the Science Citation Index that the institute had been sending to the Library of the Hungarian Academy of Sciences had been confiscated because ISI had no export license for Hungary. The tapes had been shipped for many years to Hungary and other Eastern bloc countries. ISI was informed that it would be able to obtain a license for Hungary but not for Poland or the Soviet Union. The reason given for the distinction pertained to the technology of the computer tape itself rather than the information contained on it.
* The ISI situation is illustrative of a broader effort undertaken by U.S. Customs Service under the code name "Operation Exodus." This program has involved surprise inspections of cargo bound for Eastern Europe and the Soviet Union and special searches of the personal effects of foreign nationals as they leave the United States. One example of the latter was an incident on May 6, 1982, during which Chinese graduate students waiting to board a flight in New York were detained and searched. Both they and their baggage were examined, but apparently nothing of a sensitive nature apparently was found.
* Two papers scheduled to be discussed at the May 1982 meeting of the Electrochemical Society in Toronto, Canada, were withdrawn. The papers, which dealt with VHSIC research, were judged to contain information too sensitive to be imparted to foreign nationals.
* The most recent case (to date) involving efforts to prevent the oral dissemination of unclassified research results at an international symposium occurred just as the Panel was concluding its

deliberations. Many researchers attending the 26th annual international technical symposium of the Society of Photo-Optical Instrumentation Engineers, held in San Diego, California, August 23-27, 1982, were informed with less than ten days, notice that the public presentation of their papers was being blocked by the Department of Defense because of national security considerations and the presence of Soviet scientists and other foreign nationals. In all over 150 of a total of 626 papers ultimately were withdrawn.

Annex

COCOM'S PROCEDURES

Membership

Japan and all of the NATO countries except Iceland are members of COCOM. Thus, there are several important sources of technology, among them Sweden, Switzerland, South Korea, and Taiwan, that are not members.

Target Countries

COCOM controls apply to the U.S.S.R., the PRC, Eastern Europe except Yugoslavia, and Asian Communist countries. Cuba is not subject to COCOM controls.

Operating Principles

Informal Basis

COCOM is not based on international treaty or law but on a gentleman's agreement. This has several important consequences for the organization's operations and effectiveness. First, COCOM's decisions are not legally binding on its members. Rather, they are recommendations, which the members must then implement through their own national laws. In the U.S. case, participation is effected through the Export Administration Act of 1979 as amended in 1981, which supersedes and incorporates the relevant provisions of the Battle Act. Second, COCOM has no enforcement mechanism or sanctions that can be brought to bear on a member that disregards its recommendations. Nonetheless, members

---

An extended version of this annex is printed with the collected working papers of the Panel, available separately. It appeared originally as an appendix to a paper entitled "Economic Interchange with the U.S.S.R. in the 1980s," which was prepared by John P. Hardt and Kate S. Tomlinson of the Congressional Research Service, Library of Congress. The Panel is grateful for their permission to reprint parts of it here.

(with exceptions that will be explored below) seem to regard COCOM decisions as obligations to be met.

Unanimity

As befits an informal organization, unanimity or, in some cases, unanimity of all members present, is the decision-making rule. This has several important consequences for COCOM's operations. First, no one member can impose its will on the others, but, paradoxically, each member has an effective veto. Secondly, COCOM's method of decision making may therefore be characterized as consensual or, in the view of some, as a search for the least common denominator.

Secrecy

Deliberations within COCOM and most of the details about its operations are not publicized. The high degree of discretion with which COCOM operates is not surprising, considering its hazy status in law, but there are other reasons for it. For some countries participation in COCOM may be incompatible with domestic law or may arouse criticisms from nongovernmetal leftist political parties.

The Lists

Description

Officially, the three lists of embargoed commodities, which are the basis for the control system, are classified, but it is possible to get a fair idea of what they contain. For example, it is well known and officially acknowledged that the items on the COCOM lists are on the U.S. Commodity Control List (CCL). Despite the fact that the COCOM lists can be partially reconstructed from the U.S. list and those of some of the other members, a commonly advanced rationale for keeping the COCOM lists secret is that publication could show the Soviet Union where to focus its R&D efforts.

The three COCOM lists are the following: (1) a munitions list, (2) an atomic energy list including all sources of fissionable materials, nuclear reactors, and reactor components, and (3) an industrial/commercial list, which includes dual-use items with both civilian and military uses. Understandings about COCOM procedures and operations are appended to the lists as footnotes. Since it is fairly clear what items belong on the munitions and atomic energy lists and because of the obvious security implications of exporting these kinds of commodities, the first two lists cause few disagreements within COCOM. As might be expected from the very nature of the commodities on it, the industrial/commercial list gives rise to most of the controversy within COCOM and accounts for most of its work. It is divided into a number of categories, according to product. According to some sources, the industrial/commercial list is divided into three sublists, depending on the degree of control.

Appendix C

A STUDY OF THE RESPONSES OF INDUSTRY TO A LETTER OF INQUIRY FROM THE NAS PANEL ON SCIENTIFIC COMMUNICATION AND NATIONAL SECURITY

Edward L. Ginzton

INTRODUCTION

Dale R. Corson, Chairman of the National Academy of Science's Panel on Scientific Communication and National Security, wrote to several leaders of large American companies seeking views from each company's vantage point on the general topic of interest. It is the purpose of this paper to summarize their general views and their responses to the six questions below:

1. To what extent is unwanted technology transfer via scientific communication a problem in your sector of industry?
2. If it is significant, what are the critical technologies and/or stages of research, development and production that are vulnerable?
3. What would be the impact of more stringent government controls on the affected industries?
4. How would the overall pace of innovation and product development be affected within your industrial sector?
5. How would the U.S. competitive position in international markets be affected by more rigorous controls?
6. In comparing the possible private sector effects and the potential threat to U.S. national security, where, in your estimation, does the national interest lie?

Eleven very thoughtful and timely responses were received (1), from eight major companies, mostly multinational and high technology. One letter was from a consultant who had previously served as administrator of a highly technical government agency, NASA. The wealth of material made our task difficult, but insures a significant contribution to the Panel's considerations.

The charge by the National Academy of Sciences to the Panel is relatively straightforward and narrow:

What is the effect on national security of "technology transfer to adversary nations by means of open scientific communications, either through scientific literature or by person-to-person communications . . . "?

Although the charge is narrow, the six questions stated above served to expand the areas of consideration and none of the respondents restricted their remarks to scientific communications alone.

It was apparent that the respondents had difficulty in restricting their views to their particular company. One respondent first localized his remarks to his own company, then presented a more global view (12).

One respondent wrote, "there are always going to be gray areas" (9). We found this to be correct. Differences in usage of words, particularly at interfaces, e.g., between science and technology, made it impossible for us to make a precise compilation of views. This imprecision sometimes led to a suggestion of contradiction even within a single letter.

It would have been helpful to us in analyzing the gray areas if the views could be quantified. This is seen to be very difficult, as only one of the respondents attempted a numerical estimate to support his point (6).

It is probably safe to say that the written responses reflect the individual's gestalt image of the relationship between scientific communications and national security that he had built over time as the result of his own experience, both within and outside his company. It is likely that the details of a response from other senior people within a company might differ, but probably not in any important way.

From a statistical point of view this sample is heavily biased as it is based on responses from very large companies. For this and other reasons, it is not felt that any meaningful statistical analysis can be performed, and these remarks are presented with the caveat that the conclusions may not apply to other subpopulations, e.g., smaller companies. The best that can be done is to record observations, some of which are based on views common to all participants, while others were mentioned by only one or two but yet may represent a consensus. It would be helpful to see if there is a consensus by asking the other participants for their concurrence with their peers' views.

CONTINUUM MODELS

Two of the respondents, J. F. Bucy (9) and L. Branscomb (8), presented models as an aid to their discussion. In the first of these models Bucy established a continuum ranging from _science_, on the left, through _technology_ to _products_ on the right. "Science" is defined as a systematic pursuit of knowledge. "Technology" is the application of that knowledge to the production of specific goods and services. "Products" are the result of technology, but are not technology.

The Branscomb model starts with _universities_ on the left, _industry_ in the middle, and _military_ to the right. A "university" generally deals with the generation and dissemination of fundamental knowledge derived from basic research. "Industry" develops proprietary information which may or may not be shared. The "military" deals with operational use of equipment and systems, the knowledge of which is not shared.

These two models have a commonality in that universities deal mostly with research, industry deals mostly with technology, and the military deals mostly with products. The Branscomb model groups by cultural similarity, while the Bucy model groups by kind of information. The Bucy model may be more germane but suffers from problems of abstraction, while the Branscomb model is more pragmatic because it is work-center oriented.

An attempt was made to infer and tabulate the answers to the six discrete questions. This was useful for correlations, but not for the purpose of drawing conclusions. From a careful reading of the letters, several inferences were drawn. The most striking but trivial is that the subject is complex, from which we can postulate that the solution may not be simple.

COMMON OBSERVATIONS OF THE RESPONDENTS

As we searched for commonality in the expressed views, it was not surprising to find that the strong consensus becomes clearer at the extremes of the continuum described above; therefore there are two consensuses.

The respondents either stated explicitly or implied that restrictions on scientific information would be deleterious to their companies innovative and worldwide competitive posture (2,3,4,5,6,7,8,9,10,11,12).

Nearly all of the respondents mentioned that it was appropriate and necessary to restrict military information (2,3,4,5,7,8,11,12).

Any consensus about the area in the center of the model--technology/industry--was much more difficult for us to determine, probably because the specific details of the technology become highly significant. Nevertheless, we did identify two views which seemed to be common--at least not in contention--to all respondents.

The current export control system, as it has been administrated, is considered by several of the major companies to be acceptable (2,5,7,11,12), but a move toward greater restrictions would be deleterious, even though the respondents could only sense what the new constraints might be. One felt that military security is very lax (6).

Generally the respondents felt controls should be minimal and be commensurate with the real problems (3,12). It was suggested by some that the government should analyze the Soviet need for and use of a specific technology and show that its transfer would be explicitly harmful to the national security (3,6,11). In short, to use an old expression, the rifle approach is preferred over the shotgun with its resultant scatter and harm to bystanders.

MISCELLANEOUS VIEWS OF THE RESPONDENTS

A number of important points were mentioned by only one or a few respondents; these probably are a consensus, but we could not be sure.

Two respondents mentioned that the United States no longer has overwhelming technological superiority nor a monopoly on technical information (2,3).

Ideas from universities are more rapidly used in the United States than in the U.S.S.R. Several mentioned that shrinking military lead time was the productive approach to solving the national security problem (3,7).

Direct business with the Soviet Union is not significant to those major companies surveyed (9).

No respondents mentioned China, and so it is not clear whether or not China is viewed as an adversary. Some, but not all, mentioned the Soviets and the Eastern bloc, but only one used the term "communists" (9).

Many technologies are dual-purpose, and commercial developments are commonly ahead of military (5,8).

Two respondents felt that foreign policy (economic warfare) and national security needs should not be mixed in framing restrictive controls (9,12). One respondent mentioned that the U.S. government should not protect industry from itself through restricting information flow to U.S. allies or friends (9). One mentioned that COCOM restrictions have been used for nationalistic economic advantage (7).

One respondent stressed the need to consider the Soviet reaction to any change in controls and expressed the belief that increased restrictions would hamper the West more than the Soviet Union. The increased cost to individual U.S. industries of obtaining technical information might be prohibitive, while the additional burden could easily be met by the Soviets (6).

Respondents pointed up the value to the United States of East to West information flow and recommended that it be encouraged (12).

One suggested a voluntary program be initiated for unusual technologies, like cryptology (4).

CONCLUSIONS

We find that the respondents from the eight major companies understand the need to restrict certain classes of technical information. This includes military information; application of certain technologies; and some fundamental work, e.g., cryptology.

However, their consensus is:

1. That controls on basic research would be harmful to their companies.
2. The export control regulations are workable and acceptable when administered *as they have been*.
3. Any tightening of regulations would reduce the effectiveness of the company, either by hampering its worldwide operations or by reducing its innovative and West-West competitive position vis-a-vis its non-U.S. based competition.

EDITORIAL COMMENTS

Although many ideas were presented by the respondents, it occurred to me that some concepts had been left unstated, concepts which may be useful to the Panel. For this reason I have taken the liberty of adding some editorial comments.

Smaller Companies

The companies in the sample above are all major U.S. companies that have relatively few commercial sales to the U.S.S.R., either because the products are not suitable or because export licenses have been restricted by U.S. authorities.

Some of the companies have shown interest in the U.S.S.R. market through high-level participation in organizations such as the U.S./U.S.S.R. Trade & Economic Council, a quasi-official organization, but have not devoted much effort to direct marketing.

For these companies and their Western competitors, sales to the U.S.S.R. are miniscule compared to their commercial business and, therefore, are not significant to their overall health.

This is not true for a large number of smaller high-technology companies in the United States, such as Varian, Hewlett-Packard, and other instrument companies. For these companies, the sale to the Soviet Union of an additional few units of a particular instrument is significant to that product line and, hence, to the company.

The costs of preparing information for an export license and the uncertainty in its issue have reduced sales to such a low level that it is no longer feasible for most of the smaller high technology companies to market _any_ products in the U.S.S.R. And the situation in China is approaching that condition.

The loss of sales is important, but more devastating is the loss of market share as the German, French, or Japanese competitors take the orders. Without a market share on a worldwide basis, many high technology product lines of the companies cannot survive.

Despite these differences, the smaller companies with which we have had contact would agree with the consensus and the views above, but with greater intensity.

Should Restrictions Be Decreased?

The sampling and the questions that were asked suffer from asking only "what if restrictions are increased?" and not "what if restrictions were decreased?" The possibility that a decrease in restrictions would have positive results is as a priori valid as the suggestion that an increase of restrictions is advantageous. This has to do with the complexities of enforcing restrictions, the ease of the Soviets' gaining information, the difficulty with which the United States gains information about the Soviet progress, East-West transfer, slowdown in innovative and competitive positions of U.S. companies, and international relations. Of course now that restrictions are in place, their removal should be used for a negotiating ploy.

Is the Current Export Control System Optimum?

It is difficult without further research to judge whether the current system of controls is about optimum in its benefit-to-cost ratio and, therefore, whether any change in either direction would tend to decrease this ratio. Further studies might determine whether 30 years of experience have produced a control system that is workable and socially acceptable, and whether any further tinkering with the system would only create problems. Since the system as previously administered worked, the burden should be on those who wish to tighten controls to prove that the benefits would heavily outweigh the costs.

Cooperation

It may very well be that a program of benign neglect of restrictions in certain technical areas may be cost effective. The cost of producing successful military systems may be minimized by observing Russian research results and benefiting from their experience. Once feasibility has been shown, the superior U.S. production system should be able to field equipment first. The gyrotron is a small case and the high thrust booster rocket is a large case in which this has happened.

The tactic of active cooperation has been useful in fusion research, the success of which would have large-scale economic and military advantages. High-powered lasers are a dual-use technology in which cooperation, or at least elimination of restrictions, might be most cost effective. Breeder reactors are another technology in which this concept might apply.

Lack of Information

The entire area of export control lacks quantitative information, or even validated case histories, with the result that each participant interprets it almost wholly from his own experience. This makes it difficult to select the middle ground, which clearly is the proper operating arena.

Free Market Tradition

Several of the respondents implicitly suggested that the government should rationally select those specific critical technologies for which restrictions would apply. Although not stated explicitly, this is tantamount to affirming the Western free market tradition of "freedom of action unless it is expressly forbidden," rather than the "action only if it is expressly permitted" view of the central planned countries of the Eastern bloc.

CORRESPONDENCE

(1) Due to space limitations, the actual letters are not included here. They are available for inspection, however, in the panel's office at the National Academy of Sciences.
(2) Ian M. Ross, President, Bell Laboratories.
(3) Robert D. Schmidt, Vice Chairman of the Board, Control Data.
(4) Edward E. David, Jr., President, Exxon Research and Engineering Company.
(5) Roland W. Schmitt, Vice President, Corporate Research and Development, General Electric Company.
(6) Richard L. Garwin, Watson Research Center, IBM.
(7) Lewis M. Branscomb, Office of Vice President and Chief Scientist, IBM.
(8) Same as (7).
(9) J. Fred Bucy, President, Texas Instruments.
(10) Thomas Paine, Thomas Paine Associates.
(11) W. C. Hittinger, Executive Vice President, RCA.
(12) Laurence J. Adams, Senior Vice President, Martin Marietta Corporation.

Appendix D

A BRIEF ANALYSIS OF UNIVERSITY RESEARCH AND DEVELOPMENT
EFFORTS RELATING TO NATIONAL SECURITY, 1940-1980

James R. Killian, Jr.

A decision made by the National Defense Research Committee (NDRC) in 1940 to turn to selected universities for the management of weapons research brought American universities for the first time into large-scale weapons research. Instead of turning to government laboratories or to industry for many major war research projects, NDRC made the revolutionary decision, the first in the nation's history, to ask universities to undertake both large and small war projects.

Their collective commitment to this effort, and their great success in carrying through war research and development under tight security, ushered in a new period in the relationship of the federal government to higher education. The Office of Scientific Research and Development (OSRD) entered into over 800 research contracts with nonprofit institutions, mainly universities. The total expended by OSRD in the conduct of these contracts was in excess of $330,000,000. Included in this program were projects which turned into major laboratories, such as the Applied Physics Laboratory at Johns Hopkins University (which was devoted to the development of proximity fuses), the center for rocket research at the California Institute of Technology, the atomic weapons development laboratory at Los Alamos managed by the University of California, the Radiation Laboratory at M.I.T. concentrating on short-wave radar and long-range navigation, and the radar counter-measures laboratory at Harvard University.

Other impressive aspects of this wartime university partnership with government are presented in <u>Scientists Against Time</u>, the Pullitzer Prize-winning history of OSRD by James P. Baxter, III. It is noted there that 69 different academic institutions were represented on the staff of the Radiation Laboratory, and that California, Columbia, Harvard, and the Woods Hole Oceanographic Institution operated major laboratories for the study of underwater sound and underwater explosions. Smaller projects were carried on by other institutions without the establishment of special laboratories; as, for example, work on explosives at the University of Michigan, on optics at the University of Rochester, and other notable work at Stanford University, Duke University, Rensselaer Polytechnic Institute, and the state universities of New Mexico, Texas, Iowa, Florida, and Ohio. At Massachusetts Institute of Technology, major contributions were made in the instrumentation and servo laboratories to the control of gunfire, which proved to be of decisive importance in protecting against airplane attacks on ship and land forces.

Baxter notes that the college and university at war had their dormitories and classrooms filled with Army and Navy trainees, along with reduced civilian student bodies; but, at the same time, they conducted war research of a secret character. Thus, these institutions managed the complexities of a first-rate security system, which at times involved armed guards and painstaking indoctrination. The security record of these academic institutions was admirable, proving that "intelligent and patriotic civilians, carefully indoctrinated as to the importance of security, can maintain secrecy as effectively as members of the armed services." The development of the atomic bomb and other weapons under tight security "was achieved," wrote Baxter, "not by the regimentation of science or industry but by the country where greatest pains had been taken to leave both free to make the most of the creative powers. Secrecy was maintanined without a Gestapo."

Most of the secret war research projects on campus and in big laboratories, such as the Radiation Laboratory, were brought to a rapid conclusion at the end of the war, particularly those that were secret and therefore inappropriate to the policies and freedom of academic institutions. A few major classified projects were continued after the war under university management but were conducted off campus. For example, the Applied Physics Laboratory has continued under the management of Johns Hopkins University, and, of course, the Los Alamos Laboratory has remained under the direction of the University of California, which also operates a second major classified research laboratory at Livermore. In the postwar period, a few new secret, large research projects, such as Lincoln Laboratories at M.I.T., were established by the universities, but they were mainly off campus. The scientists who worked on secret projects during the war sought with all deliberate speed to return to university environments where they and their graduate students could work in complete freedom, and practically all universities banned secret work from their campuses.

The achievements of university research during the war led the Department of Defense to fund on-campus basic research generously in the postwar period. The Office of Naval Research (ONR) moved quickly to aid universities to reestablish their graduate programs in science and technology, thereby setting a pattern of benign sponsorship that recognized those special characteristics of universities which emphasized the essential values of academic freedom and the admission (and freedom of choice) of qualified students, including foreign nationals. ONR established contracting principles and procedures that paved the way for the National Science Foundation and that were generally adopted by all parts of the Department of Defense and by other defense-related government agencies, such as the Atomic Energy Commission and its successor.

Thus, the DOD played a major part in underwriting U.S. world leadership in unclassified science and technology. There was clear recognition that this achievement was an essential contribution to national security and that the participating universities played a central role by adhering firmly to the "four freedoms" of university life. By seeking to avoid unnecessary intrusions, regulations, and secrecy requirements, the national security agencies freed the universities to work in a manner that enabled them to be most productive and creative.

Appendix E

VOLUNTARY RESTRAINTS ON RESEARCH WITH NATIONAL SECURITY IMPLICATIONS: THE CASE OF CRYPTOGRAPHY, 1975-1982

Mitchel B. Wallerstein
Staff Consultant

In 1977 the Institute of Electrical and Electronics Engineers (IEEE) scheduled a symposium at which several important papers on cryptography were to be presented. Research had established a basis for developing powerful new encryption schemes, using fundamental concepts of computer science, and examples of these schemes were included in the papers. Prior to the symposium, however, a letter arrived at IEEE headquarters warning that the presentations might subject the authors and the IEEE to prosecution under the Arms Export Control Act of 1976. The letter was signed by an IEEE member, Joseph Meyer, who gave only his home address, but who turned out to be an employee of the National Security Agency (NSA). The function of the NSA is to intercept and decipher the communications of foreign governments and to safeguard the secret communications of the United States.

After due deliberation, the IEEE decided to proceed with the symposium, although the papers of some graduate students were presented by their faculty advisors to ensure legal backing from their universities. No action was taken by the government. (It should be noted that Admiral B. R. Inman, then director of NSA, has denied that NSA attempted to suppress scholarly work in cryptography, citing a Senate committee finding that Meyer's letter to the IEEE was a personal initiative.)

Two other events also occurred in 1977. In October the University of Wisconsin at Milwaukee filed a patent application (through an affiliated foundation) for an encryption device invented by George Davida, associate professor of electrical engineering and computer science. Six months later Davida received a letter from the U.S. Patent and Trademark Office informing him that the Invention and Secrecy Act of 1951 had been invoked and that if the principles of his invention were disclosed to anyone other than federal agents, he would be subject to

---

The material in this appendix is drawn from a number of sources, including (a) verbatim portions of an article by Stephen H. Unger, "The Growing Threat of Government Secrecy," Technology Review (February/March 1982), pp. 32-33, 39, 84-85; (b) a memorandum prepared for the Panel by the Office of the Director, National Science Foundation; and (c) testimony presented to the Panel in briefings by Martin Hellman, Howard Rosenblum, and Ronald Rivest.

a $10,000 fine and two years in prison. The Patent Office did not indicate how long the invention had to be kept secret, did not justify the secrecy order, and did not indicate whether there was any way to appeal its decision.

Meanwhile, three engineers in Seattle filed an application for a patent on an inexpensive voice scrambler they planned to market. They too were the subject of secrecy order from the Patent Office. A furor arose around both cases as protests were filed and widely reported. In June 1978 the secrecy order involving Davida's invention was rescinded, and the restriction on the scrambler unit was lifted the following October.

A related sequence of events began in 1975, when a grantee of the National Science Foundation (NSF) inquired whether NSA had sole statutory authority to fund research in cryptography and whether other federal agencies were specifically enjoined from supporting that type of work. After investigation by the NSF legal staff, no basis was found for such a belief.

The matter of support for cryptography research was raised more formally in May 1977 when two NSA representatives visited the Division of Computer Research at NSF to explore ways of improving the coordination of policy between the two agencies. At that time an NSF program officer agreed to send proposals for funding research in cryptography to NSA for review, but with the caveat that an NSA recommendation against funding that gave no reasons for the recommendation would be considered unacceptable. NSF therefore reserved the right to fund such research at its own discretion. This agreement between the two agencies, confirmed in a letter to NSA from the Director of the NSF Division on Mathematical and Computer Sciences in November 1980, is now observed informally by all other NSF divisions as well.

In September 1978 NSF Director Richard Atkinson visited the NSA to discuss the likely response of NSA if NSF-supported basic research began to impinge on areas related to national security. To help prevent problems of this nature, Atkinson proposed that NSA sponsor a small unclassified research program to increase the overall level of support for cryptographic research and to differentiate between the areas to be funded by NSF and those to be funded by NSA. Meetings on such a program were never convened, but NSA subsequently established an unclassified research grants program, which made its first award in FY 1982. The NSF is cooperating in this new program and has made one joint award with the NSA.

The next development occurred in July and August 1980 when NSF received two letters from Admiral B. R. Inman, then Director of NSA, concerning research proposals submitted by Leonard Adelman and Ronald Rivest, respectively. NSA had reviewed the proposals and had decided that the probable results, if openly published, would have a serious negative impact on national security. The NSA proposed that both Adelman and Rivest contact it directly regarding support for their proposals. The NSF's response to the Inman letters was largely determined by its responsibilities under Executive Order 12065, which states that when an employee or contractor of an agency not having original classification authority originates information believed to require classification, the information must be transmitted to the relevant agency that has authority to classify that information.

The Foundation's policy was elaborated in a letter from its then Acting Director, Donald Langenberg, to *Science* and to *Nature* on November 6, 1980. The letter contained two main points: (1) the Foundation would continue to support cryptographic research while coordinating its research support with the NSA and encouraging NSA to develop its own program of support for basic research on cryptography; and (2) the Foundation would ensure that its reporting requirements were adequate to allow it to meet its responsibilities with respect to classification. The Adleman proposal was approved by the NSF on December 9, 1980, and the award letter included a statement of NSF policy and an elaboration of the reporting requirements:

> The National Science Foundation does not expect that results of basic research it supports will be classified, except in very rare instances. Further, while NSF does not have classification authority, it has the responsibility to refer any information which NSF has reason to believe might require classification to the agency with appropriate subject matter interest and original classification authority. Therefore, the grantee is responsible for immediately notifying the NSF Program Official of any data, information, or materials developed under this grant which may require classification. The grantee shall, prior to dissemination or publication of potentially classifiable research results obtained under this grant, allow NSF the option to review such materials. The grantee shall defer dissemination or publication pending the review and determination that the results are not classified, provided such review and determination are completed within sixty days of receipt by NSF of such material. If the review results in classification, the grantee agrees to cooperate with NSF or other U.S. agencies in securing all related notes and papers.

M.I.T., where Rivest worked, found the language in the Adleman award to be in conflict with its policy on cryptographic research, and unnecessary as well, since the Institute routinely send all of its cryptography-related research results to the NSA at the same time as it sends them out for technical comment from others in the field. After negotiation between M.I.T. and the NSF, mutually satisfactory wording for the reporting requirement was worked out, and a grant was made to Rivest on September 25, 1981. On April 2, 1982, President Reagan signed Executive Order 12356 on National Security Information. It states, in part:

> when an employee, contractor, licensee, or grantee of an agency that does not have original classification authority originates information believed by that person to require classification, the information shall be protected in a manner consistent with this order and its implementing directives.

The new executive order, in other words, places the responsibility for the initial judgment about the sensitivity of research results squarely on the grantee.

MAJOR GOVERNMENT CONCERNS ABOUT DISSEMINATION

The government has a number of reasons for its efforts to restrict the open dissemination of research results in cryptology. It is worried that open publication would jeopardize national security by making available to foreign governments encryption techniques that NSA would have difficulty breaking, call to the attention of foreign governments the vulnerability of their own encryption methods, and reveal knowledge that might endanger the inviolability of codes used by the United States. It is important to note that those aspects of cryptology that are applicable to national defense are considered a munition and require a license for export under the Arms Export Control Act of 1976.

The inviolability of U.S. codes is particularly important because of length of time during which codes and encrypting devices normally remain in use. NSA is now working on codes and equipment whose useful lifetime will extend through the year 2030. At the same time, however, some of the encoding equipment still in use today dates from not long after the end of World War II. Thus, if theoretical information on the design of newer encrypting equipment were to become available, the working lifetime of older machines would be reduced substantially. Finally, it is often the case that cryptographic equipment is modified incrementally in order to extend its lifetime. If state-of-the-art information were published more speedily, the practice of making incremental changes would also have to be discarded.

A concern of an entirely different sort flows from the growing dependence in the United States on electronic communications. New types of fraud have become possible, based on the manipulation of data in computer storage banks or the interception and transformation of coded information. This raises the possibility that foreign agents could cause national economic chaos by manipulating data. One defense against this kind of "data sabotage" would be the development and deployment of powerful encryption and verification systems in the business community. In this case, however, excessive secrecy in cryptological research could actually _impair_ national security.

FORMATION OF THE PUBLIC CRYPTOGRAPHY STUDY GROUP

One outgrowth of the cryptography controversy was the formation in 1980 of the Public Cryptography Study Group (PCSG), which was created by the American Council on Education and has been funded by the NSF. The nine-member group includes mathematicians and computer scientists nominated by various professional societies, university administrators, and the general counsel of NSA. The group's goal was to find a way to satisfy NSA's concerns about the publication of cryptographic research papers without unduly hampering such research or impairing First Amendment rights.

Although the PCSG initially did not wish to be bound by national security restrictions in considering various options, it ultimately agreed to accept the need for such constraints as a working hypothesis. It first considered a mandatory system, backed by the NSA, that would require all papers dealing with cryptography (as defined by the NSA) to be submitted to the Agency for prepublication review. This proposal was rejected, partly because the group felt that it had not been able to assess clearly the need for secrecy. (The PCSG neither sought nor obtained security clearance for its members during its deliberations.) The Group also felt that a voluntary arrangement would be more likely to gain the cooperation of researchers.

The PCSG eventually recommended the establishment on a trial basis of a system in which NSA would invite authors to submit cryptography manuscripts voluntarily for prior review at the same time that the manuscripts were submitted to journals. NSA would determine the research areas to be covered by the system after consultation with the appropriate technical societies. Manuscripts would be returned promptly to the authors with explanations "to the extent feasible of proposed changes, deletions, or delays in publication, if any." An author who disagreed with NSA's views on a manuscript could request a review by a committee composed of two members appointed by the director of the NSA and three appointed by the President's science advisor. The entire process would be voluntary, with neither authors nor publishers required to participate or comply with any proposed restrictions.

This proposal was accepted by all members of the PCSG except Professor Davida, who wrote a minority report arguing against any restraints. Among his many objections was the difficulty of distinguishing between basic research and knowledge directly applicable to actual systems. Davida also was concerned that a voluntary system could be a first step toward a compulsory system, and that the PCSG report could be used to support NSA's argument about the necessity of government controls over cryptographic research.

To date, approximately 46 papers have been handled using the procedures proposed by the PCSG. Only two of the papers were deemed by NSA to have implications for national security, and in both cases the problems were resolved to the satisfaction of all parties. It should be noted, however, that the NSA is not entirely satisfied with the PCSG solution. The Agency's two principal concerns are that the present arrangement sets a precedent for the future and that there is potential danger in _simultaneous_ review for security and publication purposes. The most significant problem, in NSA's view, is the possibility that a "blockbuster" paper--i.e., a paper reporting on research constituting a radical breakthrough in current knowledge--might slip through the system and seriously damage national security. The Agency believes that, as more researchers move into the field of cryptography--due, in part, to increased private sector interest--the potential for a blockbuster paper will increase significantly.

WIDER APPLICABILITY OF THE CRYPTOGRAPHY MODEL

The question of whether PCSG's solution for cryptographic research might be applied to research papers in other fields is beyond the scope of this paper. However, since cryptography has a number of characteristics that are unique, these characteristics would have to be taken into account in determining how the solution could be applied elsewhere.

First, the field of cryptography involves only a few dozen researchers--most of whom are working colleagues--and the publication of less than 100 papers per year. Second, the agency with the principal interest in cryptography, the NSA, is both technically competent and mission-oriented. In other words, it is engaged in the direct use of cryptography. Third, the frequency of problem papers--i.e., papers that would interfere with NSA's mission--is small. These characteristics do not prevail in other areas of science and technology. Hence, it is far more difficult for the government to evaluate the potential impact on national security of any single research paper in other areas of science and technology.

Appendix F

THE ROLE OF FOREIGN NATIONALS STUDYING OR WORKING IN U.S. UNIVERSITIES AND OTHER SECTORS

Mitchel B. Wallerstein
Staff Consultant

The number of foreign students in higher education in the United States increased substantially during the 1970s, especially at the graduate level. Underlying this trend were two major factors: (1) an increased demand for U.S. training to meet the needs of foreign nations for skilled scientific and engineering personnel, and (2) increased recruitment of foreign students by U.S. institutions to augment domestic enrollment. This trend is indicated clearly in Table 1. Enrollment of foreign students doubled during the 1970s at both the undergraduate and graduate levels, rising to almost 290,000 students in 1979.

GRADUATE TRAINING

The proportion of full-time graduate students in science and engineering (S&E) who were from foreign countries rose from 16 percent in 1974 to 20 percent in 1979. Although there was an increase in almost all S&E fields between 1974 and 1979, the growth was most dramatic in engineering and mathematical/computer sciences. Over 40 percent (16,200) of the 1979 graduate enrollment in engineering, and over 30 percent (4,300) of the enrollment in mathematical/computer sciences, consisted of foreign students (see Figure 1).

DOCTORATE PRODUCTION

Approximately 3,600 (or 1 out of every 5) S&E doctorates granted by U.S. universities in 1979 were awarded to foreign citizens. This is

---

The material in this appendix was derived primarily from two sources: (1) a National Science Foundation report, "Foreign participation in U.S. science and engineering higher education and labor markets" (NSF 81-316), and (2) data provided by the Commission on Human Resources of the National Research Council. Due to general nature of the data, however, it was not possible to determine the number of foreign nationals from any particular country participating in U.S. scientific and technological enterprises.

TABLE 1 Foreign Enrollment in US Institutions of Higher Education, Selected Years, 1954-1979

	All Institutions		
Selected Years	Total Enrollment	Foreign Enrollment	Foreign as a Percentage of Total
1954	2,499,800	34,200	1.4
1964	5,320,000	82,000	1.5
1970	8,649,400	144,700	1.7
1975	11,290,700	179,300	1.6
1976	11,121,400	203,100	1.8
1977	11,415,000	235,500	2.1
1978	11,392,000	263,900	2.3
1979	11,707,100	286,300[a]	2.4[a]

[a]Preliminary.

SOURCES: National Center for Education Statistics and Institute of International Education.

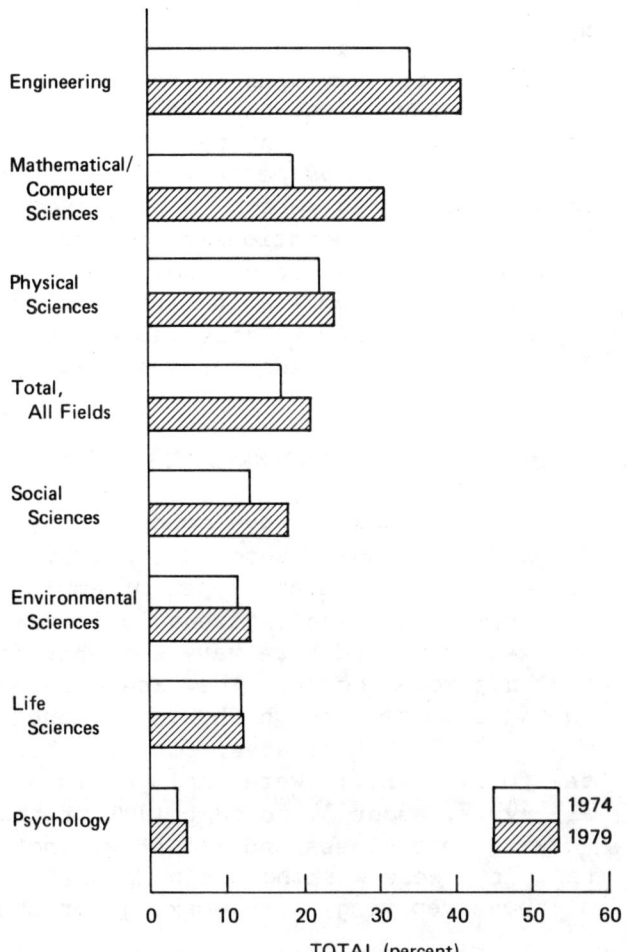

FIGURE 1 Foreign students as a percentage of full-time graduate science/engineering enrollment in doctorate-granting institutions within fields, 1974-1979.

SOURCE: National Science Foundation.

broken down by field of study in Table 2. In engineering alone, foreign nationals obtained about 1,200--or almost half--of the doctorates received by graduate students. The share of all S&E doctorates awarded to foreign nationals increased from about 15 percent in 1960 to 21 percent in 1979. Since then, it has remained relatively stable.

The large number of foreign citizens obtaining doctorates in the United States indicates the large amount of foreign interest in the high technology work underway in U.S. institutions. Foreign nationals with temporary visas received at least 20 percent of the S&E doctorates awarded in 1979 in each of more than 40 subspecialties. The largest percentages were in fuel technology/petroleum engineering (76 percent) and agricultural engineering (50 percent).

Table 3 lists the ten leading U.S. institutions that granted doctorates to foreign full-time graduate students. The University of California-Berkeley and M.I.T. were first and second, respectively, in both 1974 and 1979, but the other institutions on the list varied substantially during the five-year period. The top ten institutions (by size) accounted for 21 percent of all foreign graduate students in 1979, compared with 23 percent in 1974.

POSTDOCTORATES

Foreigners constituted about one-third (or almost 6,500) of the S&E postdoctorates employed in doctorate-granting institutions in 1979, down from almost one-half in 1977. Figure 2 shows that two of every three postdoctorate engineers in 1979 were foreign nationals. Similarly, about 50 percent of the postdoctorate positions in the physical sciences were held by persons with foreign citizenship. Likewise, foreign nationals held about 45 percent of the postdoctorate positions in mathematical/computer sciences. Table 4 presents a summary of the ten leading institutions for foreign S&E postdoctorate employment in 1979. These universities accounted for 27 percent of the foreigners with postdoctorates working in U.S. institutions.

CHARACTERISTICS OF EMPLOYMENT DISTRIBUTION

Data from annual surveys conducted by the National Research Council's Commission on Human Resources paints a more detailed picture of the fields of interest and types of employment of "science/engineering Ph.D.s with foreign citizenship in the United States in 1981." The figures presented here vary somewhat from the preceding data because they are more recent. They are also particularly noteworthy in that they encompass foreign Ph.D.s in noneducational areas of S&E employment. Table 5 indicates, for example, that while about 55 percent of the foreign Ph.D.s were employed in U.S. educational institutions of all types, about 37 percent (391 of those replying to the survey) were working in business and industry, another 2.3 percent (42 of those replying) were working for nonprofit organizations, and 1.6 percent (22 of those replying) were working for the U.S. government. Foreign

TABLE 2 Number and Percent Distribution of Ph.D. Recipients by Type of Citizenship for Selected Years

Field	1960	1965	1970	1975	1979
Total Science/Engineering					
Number of Ph.D.s Awarded	6,300	10,500	17,800	18,500	17,200
Percent US	85	83	82	78	79
Percent Foreign	15	17	18	22	21
Permanent Residents (immigrants)	(3)	(3)	(6)	(7)	(6)
Temporary Residents (nonimmigrants)	(12)	(14)	(12)	(15)	(15)
Physical Sciences					
Number of Ph.D.s Awarded	1,900	2,900	4,400	3,600	3,300
Percent US	88	86	84	77	79
Percent Foreign	13	15	16	23	21
Permanent Residents (immigrants)	(3)	(3)	(6)	(8)	(6)
Temporary Residents (nonimmigrants)	(10)	(12)	(10)	(15)	(15)
Mathematical Sciences					
Number of Ph.D.s Awarded	300	700	1,200	1,100	1,000
Percent US	81	86	84	76	74
Percent Foreign	19	14	16	24	26
Permanent Residents (immigrants)	(4)	(3)	(5)	(7)	(7)
Temporary Residents (nonimmigrants)	(15)	(11)	(11)	(17)	(19)
Engineering					
Number of Ph.D.s Awarded	800	2,100	3,400	3,000	2,500
Percent US	77	79	75	58	53
Percent Foreign	23	22	26	42	47
Permanent Residents (immigrants)	(7)	(6)	(12)	(14)	(13)
Temporary Residents (nonimmigrants)	(16)	(16)	(14)	(28)	(34)
Agriculture					
Number of Ph.D.s Awarded	400	600	800	900	900
Percent US	74	67	70	63	65
Percent Foreign	26	33	30	37	35
Permanent Residents (immigrants)	(4)	(3)	(5)	(8)	(3)
Temporary Residents (nonimmigrants)	(22)	(30)	(25)	(29)	(32)
Life Sciences (excl. Agric.)					
Number of Ph.D.s Awarded	1,200	2,000	3,400	3,600	3,600
Percent US	85	81	84	85	88
Percent Foreign	15	19	16	15	12
Permanent Residents (immigrants)	(3)	(3)	(4)	(6)	(4)
Temporary Residents (nonimmigrants)	(12)	(16)	(12)	(9)	(8)
Social Sciences					
Number of Ph.D.s Awarded	1,700	2,400	4,600	6,200	5,900
Percent US	88	87	86	86	87
Percent Foreign	12	13	14	14	13
Permanent Residents (immigrants)	(3)	(3)	(5)	(4)	(3)
Temporary Residents (nonimmigrants)	(9)	(10)	(9)	(10)	(10)

Note: Percents calculated from unrounded numbers. Detail may not add to total due to rounding.

SOURCE: National Science Foundation and National Research Council, unpublished tabulations.

TABLE 3 Ten Leading Doctorate-Granting Institutions in Foreign Full-Time Graduate Science/Engineering Enrollment, 1979 and 1974

Institution	Rank 1979	Rank 1974	Number 1979	Number 1974	Percent Change 1974-79
Total, all institutions	–	–	44,800	31,700	41
Total, leading 10 institutions	–	–	9,170	7,090	29
University of California-Berkeley	1	1	1,239	1,201	3
Massachusetts Institute of Technology	2	2	1,101	881	25
Ohio State University	3	8	1,002	610	64
University of Wisconsin-Madison	4	3	904	750	21
University of Michigan	5	9	874	600	46
University of Illinois-Urbana	6	7	864	686	26
Stanford University	7	4	861	725	19
University of California-Los Angeles	8	13	830	467	78
University of Southern California	9	15	774	451	72
Cornell University	10	5	722	711	2
All other institutions			35,620	24,610	45

SOURCE: National Science Foundation.

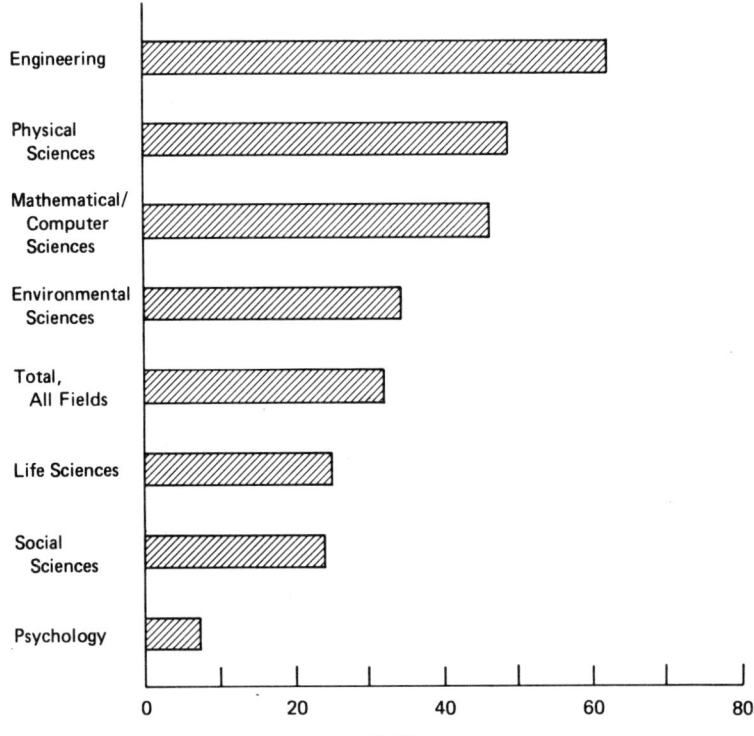

FIGURE 2 Foreign students as a percentage of total science/engineering postdoctorates in doctorate-grating institutions within fields: 1979.

SOURCE: National Science Foundation.

TABLE 4 Ten Leading Institutions in Foreign Science/Engineering Postdoctorate Employment, 1979

Institution	Rank	Number	Percent Distribution
Total, all institutions	–	6,080	100
Total, leading 10 institutions		1,649	27
Harvard University	1	297	5
University of California-Berkeley	2	189	3
Stanford University	3	171	3
Massachusetts Institute of Technology	4	160	3
University of Wisconsin-Madison	5	158	3
University of California-Los Angeles	6	145	2
University of Southern California	7	143	2
Cornell University	8	139	2
University of California-San Francisco	9	127	2
Yale University	10	120	2
All other institutions		4,430	73

SOURCE: National Science Foundation.

TABLE 5 Number of Employed Science/Engineering Ph.D.s with Foreign Citizenship in the United States in 1981 by Field of Doctorate and Type of Employer

1982 Type of Employer		All Fields	Field of Doctorate					
			Mathematics	Computer Science	Physics	Chemistry	Engineering	Bioscience
Employed Population[a]	N	1,328	107	25	146	187	181	208
	V%[b]	100.0	100.0	100.0	100.0	100.0	100.0	100.0
Educational Institution	N	800	84	13	84	91	57	160
	V%	54.5	78.3	43.5	58.7	45.5	24.5	76.2
4-Year College/University/Medical School	N	789	82	13	83	91	57	159
	V%	54.0	76.1	43.5	58.6	45.5	24.5	76.0
Business/Industry[c]	N	391	19	12	47	85	116	26
	V%	36.8	18.2	56.5	32.0	51.7	71.6	10.1
U.S. Government	N	22	1	–	3	1	–	4
	V%	1.6	0.3	–	2.6	0.9	–	4.1
Other Nonprofit Organization	N	42	3	–	8	4	4	4
	V%	2.3	3.2	–	4.8	1.1	1.8	2.5

NOTE: In view of the lack of a comprehensive sampling frame for foreign-earned doctorates in the United States, few additions of segment seven cases have been made to the sample since the 1973 survey. Therefore, the number of science and engineering Ph.D.s who are foreign citizens may be somewhat underestimated.
[a]Includes those individuals who were full-time employed, part-time employed, or on postdoctoral appointments.
[b]V% = Vertical percentage.
[c]Includes those self-employed.

SOURCE: 1981 Survey of Doctorate Recipients, National Research Council.

TABLE 6 Number of Employed Science/Engineering Ph.D.s with Foreign Citizenship in the United States in 1981 by Selected Field of Doctorate and Primary Work Activity

Primary Work Activity		All Fields[a]	Field of Doctorate					
			Mathematics	Computer Science	Physics	Chemistry	Engineering	Bioscience
Total[b]	N	1,328	107	25	146	187	181	208
	V%[c]	100.0	100.0	100.0	100.0	100.0	100.0	100.0
Research and Development	N	675	33	14	99	121	99	130
	V%	52.8	29.2	62.5	65.1	72.0	58.6	66.9
Basic Research	N	364	21	6	55	62	10	114
	V%	24.6	20.2	32.9	35.8	33.7	6.0	59.7
Applied Research	N	226	8	5	25	49	55	13
	V%	18.7	6.6	16.3	16.0	32.1	28.4	6.4
Development/Design	N	85	4	3	19	10	34	3
	V%	9.4	2.4	14.4	13.4	6.2	24.2	0.8
Management/Administration	N	115	4	4	10	19	20	15
	V%	10.3	3.5	8.8	6.3	9.6	12.7	6.7
Management of R&D	N	74	1	4	8	14	17	10
	V%	6.7	0.3	8.8	6.1	6.7	12.0	4.9
Teaching	N	375	59	7	30	31	34	41
	V%	25.3	54.3	28.7	22.3	10.9	14.5	17.4
Consulting/Professional Services	N	84	7	–	4	3	19	12
	V%	6.3	7.5	–	3.9	1.3	11.1	5.0

NOTE: In view of the lack of a comprehensive sampling frame for foreign-earned doctorates in the United States, few additions of segment seven cases have been made to the sample since the 1973 survey. Therefore, the number of science and engineering Ph.D.s who are foreign citizens may be somewhat underestimated.
[a]These figures represent more than the totals presented in the table.
[b]Includes those individuals who were full-time employed, part-time employed, or on postdoctoral appointments.
[c]V% = Vertical percentage.

SOURCE: 1981 Survey of Doctorate Recipients, National Research Council.

nationals in business and industry were heavily represented in the fields of engineering, chemistry, and physics, but were found in other fields as well. Most of those working for the U.S. government were employed in bioscience, physics, or chemistry.

In Table 6 the doctoral fields of foreign national Ph.D.s and their primary work activity are shown. Approximately 53 percent were engaged in R&D, with lesser numbers in R&D management, consulting, or teaching. The heaviest concentration of foreign national Ph.D.s working in R&D was in chemistry, followed closely by bioscience, physics, computer science, and engineering. With the exception of engineering, a larger proportion of the researchers were engaged in basic research than in applied science or development/design.

Table 7 shows foreign national Ph.D.s by type of employer and primary work activity. Here again the majority (52.8 percent, or 678, of those replying to the survey) were engaged in R&D, primarily within educational institutions (45.6 percent, or 356, of those replying) and business/industry (45.5 percent, or 250, of those replying). A majority of the foreign nationals involved in R&D in educational institutions were working in basic research, while those employed by business/industry tended to be in applied research or development/ design. The table also reveals that the federal government was

TABLE 7 Employed Science/Engineering Ph.D.s with Foreign Citizenship in the United States in 1981 by Selected Type of Employer and Primary Work Activity

			1981 Type of Employer				
Primary Work Activity		Total Employed[a]	Educational Institution Total	College/ University/ Medical School	Business/ Industry[b]	U.S. Government	Other Nonprofit Organization
Total[c]	N	1,328	800	789	391	22	42
	H%[d]	100.0	54.0	54.0	36.8	1.6	2.3
	V%[d]	100.0	100.0	100.0	100.0	100.0	100.0
Research and Development	N	675	356	355	250	16	24
	H%	100.0	45.6	45.6	45.5	2.3	2.8
	V%	52.8	44.2	44.5	65.2	77.4	64.7
Basic Research	N	364	280	280	37	13	18
	H%	100.0	77.1	77.1	11.7	3.9	4.4
	V%	24.6	34.9	35.2	7.9	62.1	46.4
Applied Research	N	226	71	70	136	3	5
	H%	100.0	26.7	26.6	63.7	1.3	2.1
	V%	18.7	9.2	9.2	32.4	15.3	16.8
Development Design	N	85	5	5	77	–	1
	H%	100.0	0.8	0.8	97.4	–	0.4
	V%	9.4	0.1	0.1	25.0	–	1.4
Management/ Administration	N	115	33	33	60	3	9
	H%	100.0	31.8	31.8	56.2	2.9	4.3
	V%	10.3	6.0	6.0	15.7	19.3	19.1
Management of R&D	N	74	12	12	51	2	6
	H%	100.0	10.9	10.9	78.6	4.4	4.6
	V%	6.7	1.3	1.3	14.2	18.7	13.3
Consulting/Professional Services	N	84	12	11	48	1	5
	H%	100.0	9.6	8.9	71.2	0.5	1.9
	V%	6.3	1.1	1.0	12.2	2.1	5.1

NOTE: In view of the lack of a comprehensive sampling frame for foreign-earned doctorates in the United States, few additions of segment seven cases have been made to the sample since the 1973 survey. Therefore, the number of science and engineering Ph.D.s who are foreign citizens may be somewhat underestimated.
[a]These figures may represent more than the totals presented in the table.
[b]Includes those self-employed.
[c]Includes those individuals who were full-time employed, part-time employed, or on postdoctoral appointments.
[d]H% = Horizontal percentage. V% = Vertical percentage.

SOURCE: 1981 Survey of Doctorate Recipients, National Research Council.

design. The table also reveals that the federal government was relatively insignificant as an employer of foreign national Ph.D.s, either in R&D or in the management of science and technology.

The data in Tables 8 and 9 provide a very limited indication of the involvement of foreign nationals in some of the sectors of the U.S. economy that are considered vital to U.S. national security--i.e., defense, space, and energy. In Table 8 the individual's generic area of interest is arrayed against his/her type of employment. Those working in the defense sector accounted for only 2.9 percent (29 of those replying to the survey) of the total foreign national workforce of Ph.D.s. This was similar to the percentage of those employed in space science (2.8 percent of those replying). Within the defense sector, Ph.D.s with foreign citizenship were found exclusively in

TABLE 8 Employed Science/Engineering Ph.D.s with Foreign Citizenship in the United States in 1981 by Selected Type of Employer and Area of National Interest

			1981 Type of Employer				
Area of Interest		Total Employed[a]	Educational Institution Total	College/ University/ Medical School	Business/ Industry[b]	U.S. Government	Other Nonprofit Organization
Total Employed[c]	N	1,328	800	789	391	22	42
	H%[d]	100.0	54.5	54.0	36.8	1.6	2.3
	V%[d]	100.0	100.0	100.0	100.0	100.0	100.0
Education	N	64	53	49	5	1	4
(Not Teaching)	H%	100.0	86.0	80.6	6.2	2.8	4.5
	V%	4.3	6.8	6.4	0.7	7.6	8.4
Health	N	273	193	193	37	5	7
	H%	100.0	69.0	69.0	16.8	2.8	1.3
	V%	16.0	20.3	20.4	7.3	28.7	9.2
Defense	N	29	12	12	17	–	–
	H%	100.0	38.3	38.3	61.7	–	–
	V%	2.9	2.0	2.0	4.8	–	–
Space	N	33	17	17	10	5	1
	H%	100.0	53.6	53.6	31.9	14.2	0.3
	V%	2.8	2.7	2.8	2.4	25.4	0.4
Energy or Fuel	N	119	49	49	63	–	4
	H%	100.0	36.7	36.7	57.7	–	3.2
	V%	12.5	8.4	8.5	19.6	–	17.2

NOTE: In view of the lack of a comprehensive sampling frame for foreign-earned doctorates in the United States, few additions of segment seven cases have been made to the sample since the 1973 survey. Therefore, the number of science and engineering Ph.D.s who are foreign citizens may be somewhat underestimated.
[a]These figures may represent more than the totals presented in the table.
[b]Includes those self-employed.
[c]Includes those individuals who were full-time employed, part-time employed, or on postdoctoral appointments.
[d]H% = Horizontal percentage. V% = Vertical percentage.

SOURCE: 1981 Survey of Doctorate Recipients, National Research Council.

educational institutions or business/industry, similar to the pattern in space research. Finally, Table 9 shows that a majority (58.4 percent or 18) of the foreign nationals working in the defense sector were involved in R&D, primarily applied research and development/design. More than one-third were working in educational areas that are related in some way to national defense.

In sum, the data on foreign national Ph.D.s in science and engineering contain few surprises. Within certain fields, foreign citizens do constitute a significant percentage of those engaged in R&D activities in both the university and industrial sectors. At the same time, however, their presence remains inconsequential in generic areas, such as defense, which are directly related to U.S. national security. It is unfortunate that more specific data, indicating country or national origin, remain unavailable, since they would probably reveal that most foreign scientists and engineers are citizens of nonadversary countries. On the basis of the evidence presented here, however, it is apparent that the total number of foreign nationals presently in the United States is significant and, most likely, still increasing.

TABLE 9 Employed Science/Engineering Ph.D.s with Foreign Citizenship in the United States in 1981 by Selected Primary Work Activity and Area of National Interest

Area of Interest		Total[a]	Research and Development				Management			Consulting/ Professional Services
			Area Total	Basic Research	Applied Research	Development/ Design	Area Total	of R&D	Teaching	
Total Employed[b]	N	1,328	675	364	226	85	115	74	375	84
	H%[c]	100.0	52.8	24.6	18.7	9.4	10.3		25.3	6.3
	V%[c]	100.0	100.0	100.0	100.0	100.0	100.0	100.0	100.0	100.0
Education (Not Teaching)	N	64	17	12	3	2	7	1	30	5
	H%	100.0	26.6	24.1	1.7	0.9	18.6	1.0	42.9	7.3
	V%	4.3	2.2	4.2	0.4	0.4	7.8	0.6	7.3	5.0
Health	N	273	171	126	39	6	23	14	49	19
	H%	100.0	68.2	50.4	12.5	5.3	7.5	5.3	15.6	5.9
	V%	16.0	20.7	32.7	10.7	9.0	11.8	12.9	9.8	15.1
Defense	N	29	18	3	10	5	1	1	8	2
	H%	100.0	58.4	6.9	24.4	27.1	1.3	1.3	36.0	4.3
	V%	2.9	3.2	0.8	3.8	8.3	0.4	0.6	4.1	2.0
Space	N	33	27	17	7	3	2	2	4	—
	H%	100.0	85.7	47.8	21.0	16.9	8.2	8.2	6.1	—
	V%	2.8	4.5	5.4	3.1	5.0	2.2	3.4	0.7	—
Energy or Fuel	N	119	75	22	38	15	6	6	25	10
	H%	100.0	56.6	11.7	24.0	20.9	6.6	6.6	22.7	12.6
	V%	12.5	13.4	5.9	16.0	27.7	8.0	12.4	11.2	24.9

NOTE: In view of the lack of a comprehensive sampling frame for foreign-earned doctorates in the United States, few additions of segment seven cases have been made to the sample since the 1973 survey. Therefore, the number of science and engineering Ph.D.s who are foreign citizens may be somewhat underestimated.

[a]These figures may represent more than the totals presented in the table.
[b]Includes those individuals who were full-time employed, part-time employed, or on postdoctoral appointments.
[c]H% = Horizontal percentage. V% = Vertical percentage.

SOURCE: 1981 Survey of Doctorate Recipients, National Research Council.

Appendix G

LETTER FROM FIVE UNIVERSITY PRESIDENTS

>Stanford University
>Office of the President
>Stanford, California 94305
>February 27, 1981

The Honorable Malcolm Baldrige
Secretary of Commerce
14th Street
Washington, D.C. 20230

The Honorable Alexander M. Haig, Jr.
Secretary of State
2201 C Street, N.W.
Washington, D.C. 20520

The Honorable Caspar Weinberger
Secretary of Defense
The Pentagon
Washington, D.C. 20301

Dear Messrs. Baldrige, Haig, and Weinberger:

I am sending the attached letter on behalf of the Presidents of Cornell University, Massachusetts Institute of Technology, California Institute of Technology, the University of California, and Stanford University to convey our grave concern about attempts to extend export restrictions to American colleges and universities. We are most anxious to cooperate in the development of alternative measures to best serve the interests of American economic development and security and would be pleased to meet with you or members of your staff to explore these issues further.

>Sincerely yours,
>
>Donald Kennedy

The Honorable Malcolm Baldrige
Secretary of Commerce
14th Street
Washington, D.C. 20230

The Honorable Alexander M. Haig, Jr.
Secretary of State
2201 C Street, N.W.
Washington, D.C. 20520

The Honorable Caspar Weinberger
Secretary of Defense
The Pentagon
Washington, D.C. 20301

Dear Messrs. Baldrige, Haig, and Weinberger:

We are writing to request clarification of the applicability of certain export restrictions to teaching and research activities conducted by American universities. We are deeply concerned about recent attempts to apply to universities the International Traffic in Arms Regulations (ITAR) and the Export Administration Regulations (EAR). Examples of such efforts by government agencies include a December 12, 1980, memorandum by the Director of the Very High Speed Integrated Circuit (VHSIC) Program Office, attempts to restrict publication of unclassified university research results arising from DOD-sponsored projects, and a Department of Commerce mandate to at least one university barring certain foreign scholars from that university's sponsored research activities due to their citizenship. Unfortunately, these initiatives appear to be only the first of many such actions to follow.

The ITAR and EAR regulations have existed for a number of years, and have not until now been applied to traditional university activities. The new construction of these regulations appears to contemplate government restrictions of research publications and of discourse among scholars, as well as discrimination based on nationality in the employment of faculty and the admission of students and visiting scholars. In the broad scientific and technical areas defined in the regulations, faculty could not conduct classroom lectures when foreign students were present, engage in the exchange of information with foreign visitors, present papers or participate in discussions at symposia and conferences where foreign nationals were present, employ foreign nationals to work in their laboratories, or publish research findings in the open literature. Nor could universities, in effect, admit foreign nationals to graduate studies in those areas. Such restrictions would conflict with the fundamental precepts that define the role and operation of this nation's universities.

The regulations could be interpreted to cover instruction and research which, although potentially useful in military applications, have much broader utility in such other areas as medical systems and communication equipment. Such interpretations of the regulations, coupled with their severe criminal penalties, could have a very real and unintended chilling effect on legitimate academic exchange.

Restricting the free flow of information among scientists and engineers would alter fundamentally the system that produced the scientific and technological lead that the government is now trying to protect and leave us with nothing to protect in the very near future. The way to protect that lead is to make sure that the country's best talent is encouraged to work in the relevant areas, not to try to build a wall around past discoveries.

It should be recognized that the only realistic way to "contain" VHSIC research is to classify the whole program. In our view this would be a self-defeating effort: the science underlying high technologies cannot be put back into the bottle. Furthermore, most universities have concluded that performance of classified research is incompatible with their essential purposes. University scientists would prefer, for the most part, to change their field of interest rather than have their research and teaching so constrained. Forcing high technology research out of universities would decrease our nation's competitive position, since the research would have to be carried out more slowly and less effectively in a classified atmosphere. Moreover, we would foreclose future research directions that would be otherwise discovered by having a continuous flow of new graduates from the university programs which have been flourishing up to this point. Elimination of such teaching and research from academic laboratories would endanger the future of graduate programs in engineering, computer science, and related fields and would result in a tremendous loss of potential high technology otherwise available to American industry. The new restrictions represent the worst possible direction: they fail to protect the status quo and virtually guarantee that there will be no future.

Moreover, application of export restrictions to universities would pose significant practical difficulties. It would be virtually impossible for most universities to administer such restrictions given the necessarily decentralized and fluid nature of most campuses. Because it is so inconsistent with their character, universities are neither structured nor staffed to police the flow of legitimate visitors to a given laboratory or the dissemination of information by their faculty at international conferences, or, indeed, even in a campus classroom where foreign students happen to be present.

The December 12, 1980, memorandum mentioned earlier pertaining to the VHSIC Program assumes basic research can be differentiated from areas such as device design and fabrication techniques, process equipment, and software, for which approval of publication or presentation normally would be denied. Such distinctions are proposed to be made by government employees, using criteria of questionable reliability and suitability. There is no such easy separation in any engineering curriculum intended to be relevant to our national industrial needs and problems. Furthermore, producing graduates with no "hands-on" experience in these areas would be of little value to American high technology industries.

The proposed extension of the restrictions to university activities ought not be made without a thorough assessment of the policy implications, the necessity and prospective effectiveness of the restrictions, the extent of disruption of the established role and operations of universities, and the serious legal and constitutional questions raised.

In the interim, it might be mutually advantageous for DOD to continue (selectively and sparingly) to rely on its classified research facilities to carry out the most sensitive segments of the VHSIC program. That has been its practice in previous years, and is far preferable to the application of these restrictive and virtually unenforceable regulations to universities. For those university activities which remain unclassified, we urge the government to cease all attempts to apply the restrictions until the broader issues are resolved.

We hope that after examining this issue carefully, you will clarify what has always been our understanding--namely, that the regulations are not intended to limit academic exchange arising from unclassified research and teaching.

Sincerely yours,

Donald Kennedy
President, Stanford University

Marvin L. Goldberger
President, California Institute
   of Technology

Paul E. Gray
President, Massachusetts Institute
   of Technology

Frank H. T. Rhodes
President, Cornell University

David S. Saxon
President, University of California

cc: Richard Allen
    National Security Advisor

Appendix H

STATEMENT OF ADMIRAL B. R. INMAN FOR THE MAY 11, 1982,
SENATE GOVERNMENTAL AFFAIRS SUBCOMMITTEE ON INVESTIGATIONS
HEARING ON TECHNOLOGY TRANSFER

Thank you, Mr. Chairman, for the opportunity to appear before this Committee this morning and to continue dialogue on this most important topic. I believe that we agree that technology transfers to the Soviets and the Eastern bloc represent a very serious problem.

I would like to take this opportunity to again enter into the public record the kinds of problems we are dealing with, and the importance of the various Soviet bloc mechanisms for acquiring Western technology.

First, as we look at the militarily useful, militarily related technology which the Soviets have acquired from the West, about 70 percent of these acquisitions have been accomplished by the Soviet and East European intelligence services, using clandestine, technical, and overt collection operations. They are trying to get technologies of proven Western weapons or component designs that can be applied directly to Soviet weapons R&D and industrial needs.

The Soviets and their Warsaw Pact allies are concentrating their efforts through purchases openly and legally and, if not successful, then illegally, including espionage. The sources of this technology may be government classified or unclassified reports, private companies' "proprietary" reports, and open-source technical documents from companies and government organizations. Embargoed equipment falls into this category as well. The Soviets undertake a very thorough vacuum cleaning of anything in the public sector that will let them better target their espionage activities.

Of the remaining 20-30 percent of the acquisitions of information of military value to the Soviets, most come through legal purchases and open-source publications or from other Soviet organizations, such as the Ministry of Trade and related international bodies; only a small percentage comes from the direct technical exchanges conducted by scientists and students.

I would like to enter into the record at this time an unclassified study from the Intelligence Community perspective of our knowledge of Soviet efforts to obtain Western technology and to use it ultimately to improve their own military capabilities.

As we look out into the 1980s, where do we believe the pressure is going to come?

Future Soviet and Warsaw Pact acquisition efforts--including acquisitions by their intelligence services--are likely to concentrate on the sources of such component and manufacturing technologies, including:

• Defense contractors in the United States, Western Europe, and Japan who are the repositories of military development and manufacturing technologies.
• General producers of military-related auxiliary manufacturing equipment in the United States, Western Europe, and Japan.
• Small and medium-size firms and research centers that develop advanced component technology and designs, including advanced civil technologies with future military applications.

The task is likely to become even more difficult in the future as several trends identified in the 1970s continue into the 1980s:

• First, since the early 1970s, the Soviets and their surrogates among the East Europeans have been increasingly using their national intelligence services to acquire Western civilian technologies--for example, automobile, energy, chemicals, and even consumer electronics.
• Second, since the mid-1970s, Soviet and East European intelligence services have been emphasizing the collection of manufacturing-related technology, in addition to weapons technology.
• Third, since the late 1970s, there has been increased emphasis by these intelligence services on the acquisition of new Western technologies emerging from universities and research centers.

The combined effect of these trends is a heavy focus by Soviet bloc intelligence on the commercial sectors in the West--sectors that are not normally protected from hostile intelligence services. In addition, the security provided by commercial firms is no match for the human penetration operations of such foreign intelligence services. But the most alarming aspect of this commercial focus by Soviet bloc intelligence services is that as a result of these operations the Soviets have gained, and continue to gain, access to those advanced technologies that are likely to be used by the West in its own future weapons systems.

I can only conclude that Western security services will be severely tested by the Soviet intelligence services and their surrogates among the East European intelligence services during the 1980s. In response, the U.S. and its Western allies will need to organize more effectively than it has in the past to protect its military, industrial, commercial and scientific communities.

I am pleased to say that coordination within the intelligence community and intelligence support to the Executive Branch departments and agencies regarding the issue of technology transfer is much better than a year ago when Bill Casey pointed out a number of deficiencies in this area to the Senate Select Committee on Intelligence. For example:

• The DCI has established a Technology Transfer Intelligence Committee (TTIC) to serve as a focal point within the intelligence community on all technology transfer issues. The Committee is able to draw on the highly skilled S&T analysts who are located throughout the military technical intelligence centers and elsewhere in the intelligence community to address this complex problem. The Committee also ensures that intelligence information collected on technology transfer is consistent with the DCI's priorities and guidance and meets the needs of community production organizations. A TTIC Subcommittee on Exchanges advises appropriate U.S. government departments and agencies of the technology transfer implications and foreign intelligence equities involved in exchange programs and commercial contacts with nationals from designated foreign countries and recommends changes as appropriate. A Subcommittee on Export Control has recently been established to provide foreign intelligence support on export control issues to appropriate U.S. government agencies.

• The intelligence agencies are now better organized to support the functions of the export control enforcement agencies. Assistant Attorney General Lowell Jensen is heading an interagency committee at Justice on export control enforcement. This group has the potential to become the most significant forum for coordinating enforcement and investigative efforts dealing with export control matters. As members of this Committee, we will ensure that it draws effectively upon appropriate intelligence data bases and support. The intelligence agencies will also become directly acquainted with the current state of the enforcement effort and the intelligence needs of the enforcement agencies but also will be in a position to acquire first hand and peruse significant information being developed by the enforcement agencies that will add to and enhance the effectiveness of the intelligence effort in the long run. Any intelligence issues that are developed in this forum may be brought back to the TTIC for appropriate consideration in an Intelligence Community setting.

• The NSC Technology Transfer Coordinating Committee, chaired by Dr. Gus Weiss, serves as a valuable high-level forum for national policy assessment and developments. It is here that the political, foreign policy, intelligence, and enforcement elements are woven together, and decisions on jurisdictional issues or program choices may be sought. Substantial intelligence support to this group will result in better understanding of the threat and greater support for the efforts of the intelligence and enforcement agencies, and will result in more considered policy determinations.

• The intelligence agencies are now in a position to make substantial contributions to the Department of Commerce Advisory Committee on Export Policy, which makes determinations concerning whether particular exports should be licensed and what general policies should be applied by the U.S.

• The Department of State's Economic Defense Advisory Committee (EDAC) Working Group II structure provides an important opportunity for intelligence, enforcement and foreign policy considerations to be discussed in the context of both general policy concerns and specific cases. Intelligence support here is essential for its value in identifying and assessing international enforcement problems and bridging the gap where there are both domestic and international aspects to a particular case.

Appendix I

EXECUTIVE ORDER ON NATIONAL SECURITY INFORMATION

NATIONAL SECURITY INFORMATION

Executive Order　　　　　　　　　　　　　　　　　　　　　**Date**
12356　　　　　　　　　　　　　　　　　　　　　　　　　　　　1982

TABLE OF CONTENTS

	Page
PREAMBLE	1
PART 1 ORIGINAL CLASSIFICATION	1
1.1 Classification Levels	1
1.2 Classification Authority	2
1.3 Classification Categories	4
1.4 Duration of Classification	5
1.5 Identification and Markings	6
1.6 Limitations on Classification	7

PART 2	DERIVATIVE CLASSIFICATION	7
2.1	Use of Derivative Classification	7
2.2	Classification Guides	8
PART 3	DECLASSIFICATION AND DOWNGRADING	8
3.1	Declassification Authority	8
3.2	Transferred Information	9
3.3	Systematic Review for Declassification	10
3.4	Mandatory Review for Declassification	10
PART 4	SAFEGUARDING	12
4.1	General Restrictions on Access	12
4.2	Special Access Programs	13
4.3	Access by Historical Researchers and Former Presidential Appointees	13
PART 5	IMPLEMENTATION AND REVIEW	14
5.1	Policy Direction	14
5.2	Information Security Oversight Office	14
5.3	General Responsibilities	16
5.4	Sanctions	16
PART 6	GENERAL PROVISIONS	17
6.1	Definitions	17
6.2	General	18

EXECUTIVE ORDER

NATIONAL SECURITY INFORMATION

This Order prescribes a uniform system for classifying, declassifying, and safeguarding national security information. It recognizes that it is essential that the public be informed concerning the activities of its Government, but that the interests of the United States and its citizens require that certain information concerning the national defense and foreign relations be protected against unauthorized disclosure. Information may not be classified under this Order unless its disclosure reasonably could be expected to cause damage to the national security.

NOW, by the authority vested in me as President by the Constitution and laws of the United States of America, it is hereby ordered as follows:

PART 1

ORIGINAL CLASSIFICATION

Section 1.1 <u>Classification Levels</u>.

(a) National security information (hereinafter "classified information") shall be classified at one of the following three levels:

(1) "Top Secret" shall be applied to information, the unauthorized disclosure of which reasonably could be expected to cause exceptionally grave damage to the national security.

(2) "Secret" shall be applied to information, the unauthorized disclosure of which reasonably could be expected to cause serious damage to the national security.

(3) "Confidential" shall be applied to information, the unauthorized disclosure of which reasonably could be expected to cause damage to the national security.

(b) Except as otherwise provided by statute, no other terms shall be used to identify classified information.

(c) If there is reasonable doubt about the need to classify information, it shall be safeguarded as if it were classified pending a determination by an original classification authority, who shall make this determination within thirty (30) days. If there is reasonable doubt about the appropriate level of classification, it shall be safeguarded at the higher level of classification pending a determination by an original classification authority, who shall make this determination within thirty (30) days.

Sec. 1.2 Classification Authority.

(a) Top Secret. The authority to classify information originally as Top Secret may be exercised only by:

(1) the President;

(2) agency heads and officials designated by the President in the Federal Register; and

(3) officials delegated this authority pursuant to Section 1.2(d).

(b) <u>Secret</u>. The authority to classify information originally as Secret may be exercised only by:

(1) agency heads and officials designated by the President in the <u>Federal Register</u>;

(2) officials with original Top Secret classification authority; and

(3) officials delegated such authority pursuant to Section 1.2(d).

(c) <u>Confidential</u>. The authority to classify information originally as Confidential may be exercised only by:

(1) agency heads and officials designated by the President in the <u>Federal Register</u>;

(2) officials with original Top Secret or Secret classification authority; and

(3) officials delegated such authority pursuant to Section 1.2(d).

(d) <u>Delegation of Original Classification Authority</u>.

(1) Delegations of original classification authority shall be limited to the minimum required to administer this Order. Agency heads are responsible for ensuring

that designated subordinate officials have a demonstrable and continuing need to exercise this authority.

(2) Original Top Secret classification authority may be delegated only by the President; an agency head or official designated pursuant to Section 1.2(a)(2); and the senior official designated under Section 5.3(a)(1), provided that official has been delegated original Top Secret classification authority by the agency head.

(3) Original Secret classification authority may be delegated only by the President; an agency head or official designated pursuant to Sections 1.2(a)(2) and 1.2(b)(1); an official with original Top Secret classification authority; and the senior official designated under Section 5.3(a)(1), provided that official has been delegated original Secret classification authority by the agency head.

(4) Original Confidential classification authority may be delegated only by the President; an agency head or official designated pursuant to Sections 1.2(a)(2), 1.2(b)(1) and 1.2(c)(1); an official with original Top Secret classification authority; and the senior official designated under Section 5.3(a)(1), provided that official has been delegated original classification authority by the agency head.

(5) Each delegation of original classification authority shall be in writing and the authority shall not be redelegated except as provided in this Order. It shall identify the official delegated the authority by name or position title. Delegated classification authority includes the authority to classify information at the level granted and lower levels of classification.

(e) _Exceptional Cases._ When an employee, contractor, licensee, or grantee of an agency that does not have original classification authority originates information believed by that person to require classification, the information shall be protected in a manner consistent with this Order and its implementing directives. The information shall be transmitted promptly as provided under this Order or its implementing directives to the agency that has appropriate subject matter interest and classification authority with respect to this information. That agency shall decide within thirty (30) days whether to classify this information. If it is not clear which agency has classification responsibility for this information, it shall be sent to the Director of the Information Security Oversight Office. The Director shall determine the agency having primary subject matter interest and forward the information, with appropriate recommendations, to that agency for a classification determination.

Sec. 1.3 _Classification Categories._

(a) Information shall be considered for classification if it concerns:

(1) military plans, weapons, or operations;

(2) the vulnerabilities or capabilities of systems, installations, projects, or plans relating to the national security;

(3) foreign government information;

(4) intelligence activities (including special activities), or intelligence sources or methods;

(5) foreign relations or foreign activities of the United States;

(6) scientific, technological, or economic matters relating to the national security;

(7) United States Government programs for safeguarding nuclear materials or facilities;

(8) cryptology;

(9) a confidential source; or

(10) other categories of information that are related to the national security and that require protection against unauthorized disclosure as determined by the President or by agency heads or other officials who have been delegated original classification authority by the President. Any determination made under this subsection shall be reported promptly to the Director of the Information Security Oversight Office.

(b) Information that is determined to concern one or more of the categories in Section 1.3(a) shall be classified when an original classification authority also determines that its unauthorized disclosure, either by itself or in the context of other information, reasonably could be expected to cause damage to the national security.

(c) Unauthorized disclosure of foreign government information, the identity of a confidential foreign source, or intelligence sources or methods is presumed to cause damage to the national security.

(d) Information classified in accordance with Section 1.3 shall not be declassified automatically as a result of any unofficial publication or inadvertent or unauthorized disclosure in the United States or abroad of identical or similar information.

Sec. 1.4 <u>Duration of Classification.</u>

(a) Information shall be classified as long as required by national security considerations. When it can be determined, a specific date or event for declassification shall be set by the original classification authority at the time the information is originally classified.

(b) Automatic declassification determinations under predecessor orders shall remain valid unless the classification is extended by an authorized official of the originating agency. These extensions may be by individual documents or categories of information. The agency shall be responsible for notifying holders of the information of such extensions.

(c) Information classified under predecessor orders and marked for declassification review shall remain classified until reviewed for declassification under the provisions of this Order.

Sec. 1.5 <u>Identification and Markings.</u>

(a) At the time of original classification, the following information shall be shown on the face of all classified documents, or clearly associated with other forms of classified information in a manner appropriate to the medium involved, unless this information itself would reveal a confidential source or relationship not otherwise evident in the document or information:

(1) one of the three classification levels defined in Section 1.1;

(2) the identity of the original classification authority if other than the person whose name appears as the approving or signing official;

(3) the agency and office of origin; and

(4) the date or event for declassification, or the notation "Originating Agency's Determination Required."

(b) Each classified document shall, by marking or other means, indicate which portions are classified, with the applicable classification level, and which portions are not classified. Agency heads may, for good cause, grant and revoke waivers of this requirement for specified classes of documents or information. The Director of the Information Security Oversight Office shall be notified of any waivers.

(c) Marking designations implementing the provisions of this Order, including abbreviations, shall conform to the standards prescribed in implementing directives issued by the Information Security Oversight Office.

(d) Foreign government information shall either retain its original classification or be assigned a United States classification that shall ensure a degree of protection at least equivalent to that required by the entity that furnished the information.

(e) Information assigned a level of classification under predecessor orders shall be considered as classified at that level of classification despite the omission of other required markings. Omitted markings may be inserted on a document by the officials specified in Section 3.1(b).

Sec. 1.6 <u>Limitations on Classification.</u>

(a) In no case shall information be classified in order to conceal violations of law, inefficiency, or administrative error; to prevent embarrassment to a person, organization, or agency; to restrain competition; or to prevent or delay the release of information that does not require protection in the interest of national security.

(b) Basic scientific research information not clearly related to the national security may not be classified.

(c) The President or an agency head or official designated under Sections 1.2(a)(2), 1.2(b)(1), or 1.2(c)(1) may reclassify information previously declassified and disclosed if it is determined in writing that (1) the information requires protection in the interest of national security; and (2) the information may reasonably be recovered. These reclassification actions shall be reported promptly to the Director of the Information Security Oversight Office.

(d) Information may be classified or reclassified after an agency has received a request for it under the Freedom of Information Act (5 U.S.C. 552) or the Privacy Act of 1974 (5 U.S.C. 552a), or the mandatory review provisions of this Order (Section 3.4) if such classification meets the requirements of this Order and is accomplished personally and on a document-by-document basis by the agency head, the deputy agency head, the senior agency official designated under Section 5.3(a)(1), or an official with original Top Secret classification authority.

PART 2

DERIVATIVE CLASSIFICATION

Sec. 2.1 <u>Use of Derivative Classification.</u>

(a) Derivative classification is (1) the determination that information is in substance the same as information currently classified, and (2) the application of the same classification markings. Persons who only reproduce, extract, or summarize classified information, or who only apply classification markings derived from source material or as directed by a classification guide, need not possess original classification authority.

(b) Persons who apply derivative classification markings shall:

(1) observe and respect original classification decisions; and

(2) carry forward to any newly created documents any assigned authorized markings. The declassification date or event that provides the longest period of classification shall be used for documents classified on the basis of multiple sources.

Sec. 2.2 <u>Classification Guides.</u>

(a) Agencies with original classification authority shall prepare classification guides to facilitate the proper and uniform derivative classification of information.

(b) Each guide shall be approved personally and in writing by an official who:

(1) has program or supervisory responsibility over the information or is the senior agency official designated under Section 5.3(a)(1); and

(2) is authorized to classify information originally at the highest level of classification prescribed in the guide.

(c) Agency heads may, for good cause, grant and revoke waivers of the requirement to prepare classification guides for specified classes of documents or information. The Director of the Information Security Oversight Office shall be notified of any waivers.

PART 3

DECLASSIFICATION AND DOWNGRADING

Sec. 3.1 <u>Declassification Authority.</u>

(a) Information shall be declassified or downgraded as soon as national security considerations permit. Agencies shall coordinate their review of classified information with other agencies that have a direct interest in the subject matter. Information that continues to meet the classification requirements prescribed by Section 1.3 despite the passage of time will continue to be protected in accordance with this Order.

(b) Information shall be declassified or downgraded by the official who authorized the original classification, if that official is still serving in the same position; the originator's successor; a supervisory official of either; or officials delegated such authority in writing by the agency head or the senior agency official designated pursuant to Section 5.3(a)(1).

(c) If the Director of the Information Security Oversight Office determines that information is classified in violation of this Order, the Director may require the information to be declassified by the agency that originated the classification. Any such decision by the Director may be appealed to the National Security Council. The information shall remain classified, pending a prompt decision on the appeal.

(d) The provisions of this Section shall also apply to agencies that, under the terms of this Order, do not have original classification authority, but that had such authority under predecessor orders.

Sec. 3.2 <u>Transferred Information.</u>

(a) In the case of classified information transferred in conjunction with a transfer of functions, and not merely for storage purposes, the receiving agency shall be deemed to be the originating agency for purposes of this Order.

(b) In the case of classified information that is not officially transferred as described in Section 3.2(a), but that originated in an agency that has ceased to exist and for which there is no successor agency, each agency in possession of such information shall be deemed to be the originating agency for purposes of this Order. Such information may be declassified or downgraded by the agency in possession after consultation with any other agency that has an interest in the subject matter of the information.

(c) Classified information accessioned into the National Archives of the United States shall be declassified or downgraded by the Archivist of the United States in accordance with this Order, the directives of the Information Security Oversight Office, and agency guidelines.

Sec. 3.3 <u>Systematic Review for Declassification.</u>

(a) The Archivist of the United States shall, in accordance with procedures and timeframes prescribed in the Information Security Oversight Office's directives implementing this Order, systematically review for declassification or downgrading (1) classified records accessioned into the National Archives of the United States, and (2) classified presidential papers or records under the Archivist's control. Such information shall be reviewed by the Archivist for declassification or downgrading in accordance with systematic review guidelines that shall be provided by the head of the agency that originated the information, or in the case of foreign government information, by the Director of the Information Security Oversight Office in consultation with interested agency heads.

(b) Agency heads may conduct internal systematic review programs for classified information originated by their agencies contained in records determined by the Archivist to be permanently valuable but that have not been accessioned into the National Archives of the United States.

(c) After consultation with affected agencies, the Secretary of Defense may establish special procedures for systematic review for declassification of classified cryptologic information, and the Director of Central Intelligence may establish special procedures for systematic review for declassification of classified information pertaining to intelligence activities (including special activities), or intelligence sources or methods.

Sec. 3.4 <u>Mandatory Review for Declassification.</u>

(a) Except as provided in Section 3.4(b), all information classified under this Order or predecessor orders shall be subject to a review for declassification by the originating agency, if:

(1) the request is made by a United States citizen or permanent resident alien, a federal agency, or a State or local government; and

(2) the request describes the document or material containing the information with sufficient specificity to enable the agency to locate it with a reasonable amount of effort.

(b) Information originated by a President, the White House Staff, by committees, commissions, or boards appointed by the President, or others specifically providing advice and counsel to a President or acting on behalf of a President is exempted from the provisions of Section 3.4(a). The Archivist of the United States shall have the authority to review, downgrade and declassify information under the control of the Administrator of General Services or the Archivist pursuant to sections 2107, 2107 note, or 2203 of title 44, United States Code. Review procedures developed by the Archivist shall provide for consultation with agencies having primary subject matter interest and shall be consistent with the provisions of applicable laws or lawful agreements that pertain to the respective presidential papers or records. Any decision by the Archivist may be appealed to the Director of the Information Security Oversight Office. Agencies with primary subject matter interest shall be notified promptly of the Director's decision on such appeals and may further appeal to the National Security Council. The information shall remain classified pending a prompt decision on the appeal.

(c) Agencies conducting a mandatory review for declassification shall declassify information no longer requiring protection under this Order. They shall release this information unless withholding is otherwise authorized under applicable law.

(d) Agency heads shall develop procedures to process requests for the mandatory review of classified information. These procedures shall apply to information classified under this or predecessor orders. They shall also provide a means for administratively appealing a denial of a mandatory review request.

(e) The Secretary of Defense shall develop special procedures for the review of cryptologic information, and the Director of Central Intelligence shall develop special procedures for the review of information pertaining to intelligence activities (including special activities), or intelligence sources or methods, after consultation with affected agencies. The Archivist shall develop special procedures for the review of information accessioned into the National Archives of the United States.

(f) In response to a request for information under the Freedom of Information Act, the Privacy Act of 1974, or the mandatory review provisions of this Order:

(1) An agency shall refuse to confirm or deny the existence or non-existence of requested information whenever the fact of its existence or non-existence is itself classifiable under this Order.

(2) When an agency receives any request for documents in its custody that were classified by another agency, it shall refer copies of the request and the requested documents to the originating agency for processing, and may, after consultation with the originating agency, inform the requester of the referral. In cases in which the originating agency determines in writing that a response under Section 3.4(f)(1) is required, the referring agency shall respond to the requester in accordance with that Section.

PART 4

SAFEGUARDING

Sec. 4.1 General Restrictions on Access.

(a) A person is eligible for access to classified information provided that a determination of trustworthiness has been made by agency heads or designated officials and provided that such access is essential to the accomplishment of lawful and authorized Government purposes.

(b) Controls shall be established by each agency to ensure that classified information is used, processed, stored, reproduced, transmitted, and destroyed only under conditions that will provide adequate protection and prevent access by unauthorized persons.

(c) Classified information shall not be disseminated outside the executive branch except under conditions that ensure that the information will be given protection equivalent to that afforded within the executive branch.

(d) Except as provided by directives issued by the President through the National Security Council, classified information originating in one agency may not be disseminated outside any other agency to which it has been made available without the consent of the originating agency. For purposes of this Section, the Department of Defense shall be considered one agency.

Sec. 4.2 <u>Special Access Programs.</u>

(a) Agency heads designated pursuant to Section 1.2(a) may create special access programs to control access, distribution, and protection of particularly sensitive information classified pursuant to this Order or predecessor orders. Such programs may be created or continued only at the written direction of these agency heads. For special access programs pertaining to intelligence activities (including special activities but not including military operational, strategic and tactical programs), or intelligence sources or methods, this function will be exercised by the Director of Central Intelligence.

(b) Each agency head shall establish and maintain a system of accounting for special access programs. The Director of the Information Security Oversight Office, consistent with the provisions of Section 5.2(b)(4), shall have non-delegable access to all such accountings.

Sec. 4.3 <u>Access by Historical Researchers and Former Presidential Appointees.</u>

(a) The requirement in Section 4.1(a) that access to classified information may be granted only as is essential to the accomplishment of authorized and lawful Government purposes may be waived as provided in Section 4.3(b) for persons who:

(1) are engaged in historical research projects, or

(2) previously have occupied policy-making positions to which they were appointed by the President.

(b) Waivers under Section 4.3(a) may be granted only if the originating agency:

(1) determines in writing that access is consistent with the interest of national security;

(2) takes appropriate steps to protect classified information from unauthorized disclosure or compromise, and ensures that the information is safeguarded in a manner consistent with this Order; and

(3) limits the access granted to former presidential appointees to items that the person originated, reviewed, signed, or received while serving as a presidential appointee.

PART 5

IMPLEMENTATION AND REVIEW

Sec. 5.1 <u>Policy Direction.</u>

(a) The National Security Council shall provide overall policy direction for the information security program.

(b) The Administrator of General Services shall be responsible for implementing and monitoring the program established pursuant to this Order. The Administrator shall delegate the implementation and monitorship functions of this program to the Director of the Information Security Oversight Office.

Sec. 5.2 <u>Information Security Oversight Office.</u>

(a) The Information Security Oversight Office shall have a full-time Director appointed by the Administrator of General Services subject to approval by the President. The Director shall have the authority to appoint a staff for the Office.

(b) The Director shall:

(1) develop, in consultation with the agencies, and promulgate, subject to the approval of the National Security Council, directives for the implementation of this Order, which shall be binding on the agencies;

(2) oversee agency actions to ensure compliance with this Order and implementing directives;

(3) review all agency implementing regulations and agency guidelines for systematic declassification review. The Director shall require any regulation or guideline to be changed if it is not consistent with this Order or implementing directives. Any such decision by the Director may be appealed to the National Security Council. The agency regulation or guideline shall remain in effect pending a prompt decision on the appeal;

(4) have the authority to conduct on-site reviews of the information security program of each agency that generates or handles classified information and to require of each agency those reports, information, and other cooperation that may be necessary to fulfill the Director's responsibilities. If these reports, inspections, or access to

specific categories of classified information would pose an exceptional national security risk, the affected agency head or the senior official designated under Section 5.3(a)(1) may deny access. The Director may appeal denials to the National Security Council. The denial of access shall remain in effect pending a prompt decision on the appeal;

(5) review requests for original classification authority from agencies or officials not granted original classification authority and, if deemed appropriate, recommend presidential approval;

(6) consider and take action on complaints and suggestions from persons within or outside the Government with respect to the administration of the information security program;

(7) have the authority to prescribe, after consultation with affected agencies, standard forms that will promote the implementation of the information security program;

(8) report at least annually to the President through the National Security Council on the implementation of this Order; and

(9) have the authority to convene and chair interagency meetings to discuss matters pertaining to the information security program.

Sec. 5.3 <u>General Responsibilities.</u>

Agencies that originate or handle classified information shall:

(a) designate a senior agency official to direct and administer its information security program, which shall include an active oversight and security education program to ensure effective implementation of this Order;

(b) promulgate implementing regulations. Any unclassified regulations that establish agency information security policy shall be published in the <u>Federal Register</u> to the extent that these regulations affect members of the public;

(c) establish procedures to prevent unnecessary access to classified information, including procedures that (i) require that a demonstrable need for access to classified information is established before initiating administrative clearance procedures, and (ii) ensure that the number of persons granted access to classified information is limited to the minimum consistent with operational and security requirements and needs; and

(d) develop special contingency plans for the protection of classified information used in or near hostile or potentially hostile areas.

Sec. 5.4 <u>Sanctions.</u>

(a) If the Director of the Information Security Oversight Office finds that a violation of this Order or its implementing directives may have occurred, the Director shall make a report to the head of the agency or to the senior official designated under Section 5.3(a)(1) so that corrective steps, if appropriate, may be taken.

(b) Officers and employees of the United States Government, and its contractors, licensees, and grantees shall be subject to appropriate sanctions if they:

(1) knowingly, willfully, or negligently disclose to unauthorized persons information properly classified under this Order or predecessor orders;

(2) knowingly and willfully classify or continue the classification of information in violation of this Order or any implementing directive; or

(3) knowingly and willfully violate any other provision of this Order or implementing directive.

(c) Sanctions may include reprimand, suspension without pay, removal, termination of classification authority, loss or denial of access to classified information, or other sanctions in accordance with applicable law and agency regulation.

(d) Each agency head or the senior official designated under Section 5.3(a)(1) shall ensure that appropriate and prompt corrective action is taken whenever a violation under Section 5.4(b) occurs. Either shall ensure that the Director of the Information Security Oversight Office is promptly notified whenever a violation under Section 5.4(b)(1) or (2) occurs.

PART 6

GENERAL PROVISIONS

Sec. 6.1 <u>Definitions</u>.

(a) "Agency" has the meaning provided at 5 U.S.C. 552(e).

(b) "Information" means any information or material, regardless of its physical form or characteristics, that is owned by, produced by or for, or is under the control of the United States Government.

(c) "National security information" means information that has been determined pursuant to this Order or any predecessor order to require protection against unauthorized disclosure and that is so designated.

(d) "Foreign government information" means:

(1) information provided by a foreign government or governments, an international organization of governments, or any element thereof with the expectation, expressed or implied, that the information, the source of the information, or both, are to be held in confidence; or

(2) information produced by the United States pursuant to or as a result of a joint arrangement with a foreign government or governments or an international organization

of governments, or any element thereof, requiring that the information, the arrangement, or both, are to be held in confidence.

(e) "National security" means the national defense or foreign relations of the United States.

(f) "Confidential source" means any individual or organization that has provided, or that may reasonably be expected to provide, information to the United States on matters pertaining to the national security with the expectation, expressed or implied, that the information or relationship, or both, be held in confidence.

(g) "Original classification" means an initial determination that information requires, in the interest of national security, protection against unauthorized disclosure, together with a classification designation signifying the level of protection required.

Sec. 6.2 General.

(a) Nothing in this Order shall supersede any requirement made by or under the Atomic Energy Act of 1954, as amended. "Restricted Data" and "Formerly Restricted Data" shall be handled, protected, classified, downgraded, and declassified in conformity with the provisions of the Atomic Energy Act of 1954, as amended, and regulations issued under that Act.

(b) The Attorney General, upon request by the head of an agency or the Director of the Information Security Oversight Office, shall render an interpretation of this Order with respect to any question arising in the course of its administration.

(c) Nothing in this Order limits the protection afforded any information by other provisions of law.

(d) Executive Order No. 12065 of June 28, 1978, as amended, is revoked as of the effective date of this Order.

(e) This Order shall become effective on August 1, 1982.

THE WHITE HOUSE

Appendix J

CORRESPONDENCE BETWEEN THE STATE DEPARTMENT AND THE UNIVERSITY OF MINNESOTA AND M.I.T. RESTRICTING FOREIGN VISITORS

This Appendix presents two case examples of the restrictions imposed by the Department of State on foreign scientists seeking to visit U.S. universities and of the university response to such restrictions. The first case concerns Qi Yulu, a scholar from the People's Republic of China interested in computer software technology, whose visit to the University of Minnesota was to be restricted. The second case involves a Soviet scientist, Mikhail Y. Gololobov, whose interests involved biological and nutritional research. Dr. Gololobov's principal placement was to be at Purdue University, with shorter visits to the Universities of New Orleans, Miami, California at Santa Cruz, and M.I.T. The correspondence reprinted here pertains to his proposed visit to M.I.T.

DEPARTMENT OF STATE
Washington, D.C. 20520
September 30, 1981

Professor W. R. Franta
Department of Computer Science
University of Minnesota
Minneapolis, MN 55455

Dear Professor Franta;

I have been unable to contact you by phone in recent days. Therefore, I am taking this opportunity to write to you concerning Qi Yulu, a Chinese scholar assigned to your department.

The U.S. government regularly reviews the programs of Chinese exchange visitors in scientific and technical programs to meet export control and national security concerns. Various government technicians have reviewed the program of Qi Yulu.

Because U.S. government policy encourages the training of accomplished Chinese scholars in modern technology and science, our concerns are primarily limited to potential transfer of classified technology or of technology which requires an export license. In cases where there appears to be a potential for some unacceptable technology loss, the Department of State often contacts the academic host for additional details about the program in question. If the concerns remain, the Department of State and the Department of Commerce inform the academic host of the pertinent export control regulations as well as requirements not to give students access to classified materials.

In the case of Qi Yulu, the reviewers' concern stems largely from concerns about the potential loss of critical U.S. technology in the area of computer software technology. This is an area with military applications. The concerns could be lessened if you could provide additional information on the planned program of study and research. As this area is also subject to export control regulations, you may be contacted by the Department of Commerce in the near future.

In the meantime, it is suggested that Qi be restricted from any access to unpublished or classified government-funded work. It is also suggested that the program emphasize course work with minimal involvement in applied research. There should be no access to the design, construction, or maintenance data relevant to individual items of computer hardware. There should be no access to source codes or their development. His access should be limited to the published software for operating systems subroutines. Within this framework, however, Qi should not be denied as full an academic program as possible.

I would take this opportunity to remind you that this office should be advised prior to any visits to any industrial or research facilities.

Mr. Keith Powell, II
September 30, 1981
Page Two

    I would appreciate hearing your reactions to this request and would appreciate your keeping us informed about the program. Do not hesitate to contact this office should you have any questions concerning this matter. Please acknowledge this letter at your earliest convenience. You may call me (202) 632-1322. I hope to hear from you soon.

                                          Sincerely Yours,

                                          Keith Powell, II
                                          Exchanges Officer

UNIVERSITY OF MINNESOTA

Office of the President
202 Morrill Hall
100 Church Street S.E.
Minneapolis, Minnesota 55455

October 16, 1981

Mr. Keith Powell, II
Exchanges Officer
Department of State
2201 C Street, N.W.
Washington, D.C. 20520

Dear Mr. Powell:

This is in reply to your letter of September 30, 1981, to Professor W. R. Franta of our Department of Computer Science, asking that the University of Minnesota impose certain restrictions on the activities of a visiting scholar from the People's Republic of China. Since the University of Minnesota currently has extensive involvement with PRC students and scholars, it seemed to me that I should be the person to respond.

Enclosed, for your background information, is a copy of the "Secrecy in Research" policy of the Regents of the University of Minnesota. You will note that we do not accept classified research, so PRC scholars will not have access to classified research on our campuses.

You should know that at least someone in the State Department must be aware of this policy, since they recently went to some length to send plant samples through a third party to one of our laboratories for testing for mycotoxins. This analysis led to the allegations that the Soviets had used biological weapons in Southeast Asia. Our faculty member was not told that the State Department was the source of the samples, and he was subsequently maligned in the press for conducting secret biological warfare research. His role has now been clarified and vindicated, but I thought you should know that the State Department has aroused considerable sensitivity on this campus. Efforts to impose vaguely defined and rather sweeping restrictions on the activities of any scholars, whether from the PRC or elsewhere, would certainly reopen these sensitivities.

I have no way to assess the State Department's review of the program of the Chinese scholar in question. I do not know who the "various government technicians" were who conducted the review or what procedures and standards they might have used. I am satisfied that the restrictions you have proposed are quite sweeping and subject to almost any interpretation that might be applied now or applied retroactively at some future date.

Mr. Keith Powell, II
October 16, 1981
Page Two

To cite a few examples, you suggested that he be restricted "from any access to unpublished or classified government-funded work." I have already pointed out that we do not have classified research; we have all kinds of unpublished government-funded research all over this campus. Your proposal would restrict him from access to all of it. You ask for course work "with minimal involvement in applied research"; I don't know what you mean by "minimal," and I have no idea how you define applied research. You ask to be informed "prior to any visits to any industrial or research facilities"; I can only interpret this to give us the choice of confining him to the student union or contacting you several times a day about his campus itinerary. Quite frankly, I find this request ironic, coming from an administration that has vowed to reduce the role of big government and eliminate unnecessary paperwork.

Both in principle and in practice, the restrictions proposed in your letter are inappropriate for an American research university. Our mission is teaching, research, and public service, and neither our faculty nor our administrators were hired to implement government security actions. We are host to a large number of Chinese scholars, we have sent several delegations to the PRC, and we have formal cooperative agreements with several educational institutions in the PRC. All these relationships were developed within our traditions of openness and academic freedom, and, to use the popular phrase, the restrictions you propose can only have a chilling effect upon the academic enterprise.

I sincerely believe that the State Department should reconsider its efforts to suggest or impose these kinds of restrictions.

                                      Cordially,

                                      C. Peter Magrath
                                      President

cc: Professor W. R. Franta
    Dean Roger Staehle
    Vice President Kenneth Keller
    Regent Wenda Moore
    Senator David Durenberger
    Senator Rudy Boschwitz
    Representative Bruce Vento
    Representative Martin Sabo

October 8, 1971

SECRECY IN RESEARCH

Article 1. The University of Minnesota shall not accept support from any source for research under a contract or a grant which would restrain the University from disclosing (1) the existence of the contract or grant; (2) the identity of the sponsor or the grantor and, if a subcontract is involved, the identity of the prime contractor if the results of the research must be reported to the sponsor, grantor, or prime contractor; and (3) the purpose and the scope of the proposed research in sufficient detail: (a) to permit informal discussion concerning the wisdom of such research within the University, and (b) to inform colleagues in immediate and related disciplines of the nature and importance of the potential contribution of the disciplines involved.

Article 2. The University of Minnesota shall not accept support from any source for research under a contract or a grant, even though it meets the requirements of Article 1, if the contract or grant limits the full and prompt public dissemination of results or specifically permits retroactive classification, except for reasons found compelling by the University community through the review process outlined in Article 4.

Article 3. The above policy shall apply to any research under a contract or grant which does not limit the full and prompt public dissemination of results at the time the research is undertaken by the University but becomes so limited thereafter. As soon as this occurs, the contract or grant, and the disposition of the results of the research obtained under such contract or grant shall be re-evaluated under the provisions of Article 4.

Article 4. a. The Director, Research Contract Coordination, or some other designated University official, shall report to the Senate Research Committee every proposed research grant or contract which meets the requirement of Article 1 but limits the full and prompt public dissemination of results. If this officer is not certain whether a particular research proposal requires the Senate Research Committee's recommendation, he shall submit the proposal to this Committee for its determination.

b. The Senate Research Committee shall recommend to the Senate acceptance or rejection of every proposed contract or grant which limits the full and prompt public dissemination of results during fall, winter, and spring quarters, in sufficient detail to permit informal discussion of the recommendation made. In addition, the Committee shall report on any problems encountered in implementing this policy.

In performing its functions hereunder, the Senate Research Committee shall be authorized to seek the advice and assistance of ad hoc subcommittees competent to pass on the particular matters that may be involved. If some other University committee also has jurisdiction in a particular case, nothing in this statement of policy shall deprive it of that jurisdiction.

c. The University Senate shall review the recommendations of the Senate Research Committee and forward its own recommendations to the President. All proposals which are to be submitted for Senate evaluation shall be accessible to members of the University community (the faculty and students) in sufficient detail to permit informed evaluation and discussion.

<u>Article 5</u>. The University shall not make available any of its facilities for which permission is required to any individual, group, or organization for research which violates this Statement of Policy. Exceptions may be made through the review procedure outlined in Article 4.

<u>Article 6</u>. The above policy shall not apply to (1) research by faculty members on leave from the University or serving as consultants, or (2) research which involves (a) the collection of confidential personal opinions and attitudes or other information pertaining to the individual persons or business entities, or (b) the analysis of the characteristics or uses of proprietary devices or substances, provided that the results of such research may be published freely in the aggregate or used to guide the design of broader research activities.

The Board of Regents approved Articles 1 and 6 of the Senate Policy on Secrecy in Research on July 10, 1969. The remainder of the policy was approved on October 8, 1971.

DEPARTMENT OF STATE

Washington, D.C. 20520

November 16, 1981

C. Peter Magrath
President
University of Minnesota
Minneapolis, MN 55455

Dear Mr. Magrath:

Thank you for your letter of October 16. You have misunderstood the content and purpose of my letter of September 30, 1981, which was neither aimed at imposing sweeping restrictions, nor at interfering with academic freedom. It was certainly not my intention to disturb sensitivities at the University of Minnesota. I do still need to hear from Professor Franta to learn additional details concerning the program of study and research for Qi Yulu.

My letter to Professor Franta was intended to be neither a definitive statement nor application of U.S. export regulations and travel controls as they apply to exchange visitors from the People's Republic of China. My intent was to encourage Professor Franta to contact me. As my letter indicated, I had been unable to reach him and hoped to converse with him in order to determine whether export administration regulations regarding the transfer of technical data would be applicable in the case of Qi Yulu. If so, then more precise guidance, tailored to Mr. Qi's program, would be provided.

There are in existence several laws, including the Arms Export Control Act and the Export Administration Act, that regulate the export of all goods, services, and technology from the United States. These acts include "technical data" as well as hardware. "Export" is defined to include the transmission or release of the relevant material to a foreign national. In general the acts exempt from control information that is already available to the public.

The applicability of the regulations varies with regard to the nationality involved. Thus, a student from the People's Republic of China could come under the purview of these regulations while a student from Canada, involved in the same program, might not. For this reason, the State Department routinely requests information on proposed programs of study for all Chinese scholars. In the vast majority of cases the information provided by the host is, in itself, sufficient to satisfy the Departments of Commerce and/or State that the aforementioned regulations would not apply.

In a few cases, when it appears that the regulations would come into force, the academic host is contacted by the State and/or Commerce

Departments. The applicable regulations are explained and suggestions are offered as to how the program could be amended to meet the requirements of law. Within that context, we then encourage the fullest academic program possible. We are always happy to provide guidance and to respond to specific questions regarding the subject areas covered by the International Traffic in Arms Regulations and the Export Administration Regulations.

Finally, with regard to travel controls, my reference was to requirements which Chinese students meet. The responsibility in this regard rests with the students themselves. They are informed of the requirements when they receive their visas in China. The enclosed memo from the International Communications Agency (ICA) explains this point as well as other matters affecting Chinese exchange visitors. This memo was sent to all universities with active exchange programs last June.

I hope the foregoing provides a clear insight into the reasons for my communication with Professor Franta. My goal was to advise him of existing laws and regulations and to learn more of Qi Yulu's program so that there would be no inadvertent violation of the law. I also intended, in the absence of a response from Professor Franta, to provide interim suggestions.

I appreciate your frank and direct response, which has led me to restructure the letter format my office has been using so as to avoid misinterpretations and to convey more clearly the nature of our concerns. I do still need appropriate information regarding Mr. Qi's program. I will again seek to contact Professor Franta in the near future. If you have additional questions or suggestions, please do not hesitate to write to me or call me at (202) 632-1322.

Sincerely yours,

Keith Powell, II
Office of Chinese Affairs

Enclosure: ICA Memo

cc: Professor W. R. Franta

UNIVERSITY OF MINNESOTA

Office of the President
202 Morrill Hall
100 Church Street S.E.
Minneapolis, Minnesota 55455

December 7, 1981

Mr. Keith Powell, II
Exchanges Officer
Department of State
2201 C Street, N.W.
Washington, D.C. 20520

Dear Mr. Powell:

I appreciate your November 16 letter and the effort it represents to avoid a needless and disruptive controversy. The University of Minnesota, both because of its international educational commitments and because of our understanding of national policy with regard to educational exchange with such significant countries as the People's Republic of China, has committed a very considerable effort to international education and, especially, the development of productive exchanges with the PRC. Neither institutionally, nor personally, do I wish to have misunderstandings or an abrasive conflict with the Department of State or the federal government on such matters.

Unfortunately, as I have reviewed your letter and comments, aided by University legal counsel, I find it impossible to alter the core position stated in my original October 16, 1981, communication.

Allow me to explain this by beginning with what appears to be the clearest issue, namely the question of travel notification. Your original letter to Professor Franta states in part, that you simply wish to ". . . remind [him] that [your] office should be advised prior to any visits to any industrial or research facilities." There is absolutely no reference to the fact that there is no obligation on Professor Franta or any other University employee to provide this information to your office. Your request for this kind of travel information, we believe, is therefore unjustified.

More generally, I hope you can understand our great caution and suspicion with regard to requests for detailed information about a scholar's course of study and research by general reference to the Arms Export Control Act and the Export Administration Act. Both of these are lengthy, complex, and ambiguous pieces of legislation and are difficult to interpret even by those dealing daily in munitions or exports. What is clear to me as an educator and a public official is my responsibility to protect the privacy of scholars at this institution under the Federal Education Rights and Privacy Act, 20 U.S.C. §1332g, as reinforced and augmented by the Minnesota Government Data Practices

Mr. Keith Powell, II
Department of State
December 7, 1981
Page Two

Act, Minn. Stat. §15.1611, et seq. If your legal department can be more specific about any authority granted to your office which supercedes my responsibilities and those of Professor Franta under the law I cite, I am willing to reexamine this position--so long as it is possible to maintain the integrity of academic principles, traditions, and obligations that are the foundation of education in a democratic society.

As your own statement indicates, the legislation you present as your authority exempts from control information available to the public. I repeat, it is the clearly stated policy of the Regents of the University of Minnesota that no classified research is to be conducted on the campus.

Under all the circumstances, I do not believe that it would be particularly helpful for Professor Franta to be in direct contact with you. Instead I suggest that any request for information on individuals be addressed to him in writing, with a copy to me, so that we can seek appropriate legal advice with regard to the University of Minnesota's responsibilities in responding. I would very much appreciate your reconsidering and reexamining the bases for requests for information under the applicable laws in view of the considerations I have noted.

To repeat, we are not interested in any confrontation or disagreeable dispute with the Department of State, and we are exceedingly anxious to maintain our international exchange activities with the PRC and other countries. But we are insistent that our faculty and our students (regardless of their country of origin) be allowed to operate in their teaching and research functions both in letter and in spirit in ways consistent with our academic traditions and the applicable laws, which we believe are totally on the side of openness and do not make it appropriate for faculty to report on or control the activities of our students.

Cordially,

C. Peter Magrath
President

CPM:kb

cc: Vice President Kenneth H. Keller, Academic Affairs
    Professor W. R. Franta, Department of Computer Science

NATIONAL ACADEMY OF SCIENCES
National Research Council
2101 Constitution Avenue, Washington, D.C. 20418 USA

COMMISSION ON INTERNATIONAL RELATIONS
Cubic Address: NARECO
TWX #: 7108 22 9589

June 1, 1981

MEMORANDUM

TO: Dr. Michael Laskowski, Jr. Purdue University, West Lafayette
Dr. G. G. Guilbault, University of New Orleans, Louisiana
Dr. Sidney W. Fox, University of Miami, Coral Gables
Dr. Nevin S. Scrimshaw, Massachusetts Institute of Technology
Dr. Anthony L. Fink, University of California, Santa Cruz

FROM: Diana B. Bieliauskas, Section on U.S.S.R. & Eastern Europe
Program Officer, 202/389-6066

SUBJECT: Proposed Scientific Exchange Visit of Dr. Mikhail Yuryevich Gololobov

The Academy of Sciences of the U.S.S.R. (ASUSSR) has proposed to the National Academy of Sciences (NAS) Dr. Mikhail Y. Gololobov, Research Fellow at the Institute of Elemento-Organic Compounds in Moscow, for a three-month visit in the U.S. within the framework of the interacademy exchange program. A copy of the official agreement governing this program is enclosed for your reference as is biographical data on Dr. Gololobov. He requests a starting date in August.

Dr. Gololobov wishes principal placement at Purdue University, to be followed by shorter visits to New Orleans, Miami, M.I.T., and Santa Cruz. As you can see from his biographical information, Dr. Gololbov is hopeful of arranging a longer stay in the U.S. than just three months. Due to budgetary considerations, however, we are not able to offer Dr. Gololobov assurance of a longer stay and will plan his program for three months for the time being.

I am writing at this time to determine if you are willing and available to receive Dr. Gololobov in accordance with this schedule. Please indicate in your response which dates you would prefer, as well as those which are impossible for you.

Dr. Gololobov
June 1, 1981
Page 3

During his stay in the U.S., Dr. Gololobov will be financially self-sufficient. Dr. Gololobov will receive a living allowance to cover food and housing. We do ask prospective hosts to assist the visiting scientists by meeting them upon arrival and locating housing for them. The NAS will pay directly for their travel, medical insurance, and any income taxes. These financial and administrative provisions are described in detail in the accompanying "Handbook for Hosts" and summarized in the "Fact Sheet."

The U.S. Department of Commerce has asked that we provide all prospective U.S. hosts of visiting scientists with information about the Export Control Act of 1949. In this regard, I call your attention to Section 16 of the Handbook.

Finally, I should add that all visits by Dr. Gololobov are subject to approval by the U.S. Department of State. Dr. Gololobov's professional program must focus on fundamental, not applied research. The State Department has informed us that:

> The subjects may not have access, whether it be visual, documentary, or verbal, to production, research, or activities funded by DOD contracts or grants.

Moreover, hosts are asked to limit the programs of visiting scientists to research that has been published in open literature. Questions of interpretation of State Department policy can be directed to Mr. William Brencick (202/632-8956). If there are any local security considerations of which you are aware that might influence your ability to receive Dr. Gololobov, please let me know.

I would be glad to discuss the proposed program of vists for Dr. Gololobov in detail with you. Please telephone me at your earliest convenience concerning your willingness to host this visiting scientist or giving suggestions for alternate affiliations.

Enclosures:
    Interacademy Agreement (NAS & ASUSSR)
    Handbook for Hosts and Fact Sheet
    Biographical data on Dr. Mikhail Y. Gololobov

cc:
    Mr. William Brencick, Department of State

DEPARTMENT OF STATE

Washington, D.C. 20520

November 6, 1981

Diana Bieliauskas
Program Officer
U.S.S.R. and Eastern Europe
National Academy of Sciences
2101 Constitution Ave.
Washington, D.C. 20418

Dear Diana,

Here are the cautions and restrictions we have placed on the Soviet candidates approved to date. If you have any concerns please call me. We await the additional material on Rudashevskiy and hope to have a determination on Ilgamov in the near future.

1. GOLOLOBOV, M.Y. - (A) During his visit to M.I.T. he should not be exposed to work there on nutritional research and possible production of food supplements. His hosts at M.I.T should be informed of the U.S. Government's concerns regarding access to genetic engineering and prevent Gololobov's access to such work at M.I.T. (B) Gololobov should have no access (visual, documentary or verbal) to production, research or other activities funded by DOD contracts or grants.

[Extraneous Material Deleted]

Sincerely,

James George Jatras
Office of Soviet Union Affairs

NATIONAL ACADEMY OF SCIENCES
National Research Council
2101 Constitution Avenue, Washington, D.C. 20418 USA

COMMISSION ON INTERNATIONAL RELATIONS
Cubic Address: NARECO
TWX #: 7108 22 9589

December 3, 1981

MEMORANDUM

TO: Dr. Michael Laskowski, Jr. Purdue University, West Lafayette
Dr. G. G. Guilbault, University of New Orleans, Louisiana
Dr. Sidney W. Fox, University of Miami, Coral Gables
Dr. Nevin S. Scrimshaw, Massachusetts Institute of Technology
Dr. Anthony L. Fink, University of California, Santa Cruz

FROM: Diana B. Bieliauskas, Section on U.S.S.R. & Eastern Europe
Program Officer, 202/334-2644, -2652

SUBJECT: Reinstatement of Proposed Scientific Exchange Visit of Dr. Mikhail Yuryevich GOLOLOBOV

Early last summer, I advised you that the Soviet Academy of Sciences had nominated Dr. Mikhail Y. Gololobov, Research Fellow at the Institute of Elemento-Organic Compounds in Moscow, for a three-month visit in the U.S. Because of budgetary restrictions, we were unable to accommodate Dr. Gololobov's visit at the time he wished to start, i.e., August 1981. Please refer to my memorandum of June 1 for further details.

Dr. Gololobov has been renominated by the ASUSSR for a three-month visit starting January 12, 1982. He will spend all of that time at Purdue University. The NAS has requested that Dr. Gololobov be nominated for a longer visit in accordance with the wishes of his primary host, Dr. Laskowski. Should Dr. Gololobov be in the U.S. for longer than three months, I will contact you again about a short visit to your institution. This would probably not take place until late spring/early summer.

For your information, Dr. Gololobov's visit in the U.S. has been cleared by the Department of State with the following restrictions:

Dr. Gololobov
December 3, 1981
Page Two

(a) During his visit to M.I.T. he should not be exposed to work there on nutritional research and possible production of food supplements. His hosts at M.I.T. should be informed of the U.S. Government's concerns regarding access to genetic engineering and prevent Gololobov's access to such work at M.I.T.

(b) Gololobov should have no access (visual, documentary or verbal) to production, research or other activities funded by DOD contracts or grants.

Enclosure:
Memorandum of June 1, 1981

cc:
Mr. James Jatras, Department of State

MASSACHUSETTS INSTITUTE OF TECHNOLOGY
International Food and Nutrition Program

20A-201
18 Vassar Street
Cambridge, Mass. 02139 U.S.A.

December 21, 1981

Francis E. Low
Provost
M.I.T., 3-208
Cambridge, MA 02139

Dear Francis:

   The import of the enclosed correspondence is so bad that I believe M.I.T. should enter some kind of formal protest. The idea that scientists from the U.S.S.R. or any other country should visit M.I.T. under the sponsorship of the Commission of International Relations of the National Academy of Sciences on the condition that they "not be exposed to work there on nutritional research and possible production of food supplements" is so outrageous as to be incredible. If you prefer that I write, please let me know.

Sincerely,

Nevin S. Scrimshaw
Institute Professor

MASSACHUSETTS INSTITUTE OF TECHNOLOGY
Office of the Provost
Cambridge, Massachusetts 02139

January 20, 1982

Dr. Frank Press
President
National Academy of Sciences
2101 Constitution Avenue
Washington, D.C. 20418

Dear Frank:

As I told you in our conversation yesterday, Professor Nevin Scrimshaw received a memoradum dated Dec. 3, 1981, from your program officer for the U.S.S.R. and Eastern Europe, relating to the visit of Dr. Gololobov from the U.S.S.R. The memorandum contained restrictions which we could neither accept nor enforce if accepted. These restrictions required that Dr. Gololobov should not be exposed to work on nutritional research, food supplement production, or any research sponsored by DOD. As I am sure you agree, these conditions are inconsistent with the spirit and practice of a university as an open community of scholars, teachers, and students.

I hope that you will be able to persuade the State Department to agree to visits whose terms make university participation possible. The exchange program is valuable, and it is important to preserve it.

Yours sincerely,

Francis E. Low
Provost

cc: Dr. N. Scrimshaw

FEL:mcs